Funny Thinking in the 21st Century

Truth or Dare?

Dr. Marcia R Pinheiro

INDEX

-

INTRODUCTION

It is pretty common for humans to present odd thinking once in a while: they look at the sea and they think that the planet is a cube and we can all die if we try to go somewhere in a boat, they look at the sky and they think someone put a cheese there, they look at a deformed person and they think they are gods, and therefore more powerful than us, etc.

Science is just a representation of what society is at a given time when let on its own. Nowadays there is a fair bit of politics involved, so that Science is more what the politicians want than what it is in fact, but, in the more remote times, Science was still an actual representation of people's thinking, like in general.

By the time of Descartes, they thought saying that thinking is existing was a big deal, so that simply stopping and imagining things was seen as a useful activity.

By the time of Zeno, one thought that it was actually possible that nobody moved, despite our impression, so that we could all have hallucinations 24/7 and then come up with funny conclusions.

By my time, Priest thought that things could be and not be at the same time in Mathematics and in the logical systems. Hyde thought that we had to put more Mathematics in our language. Corcoran thought that the mathematical induction was something that could be perfected, since perhaps half of the process would lead to the same results. Sever thought that the how did not matter in Mathematics. Shepherd thought that his eyes could see better than young eyes and better than the computers. UFRJ thought that they could bug people in their heads if what they thought was inadequate. RMIT thought that conversing before acting in a real-life situation is a waste of time, that they could simply do things to their postgraduate students.

People do have odd thinking at least sometimes.

This book is about a few of the greatest figures of the 21st century who were somehow involved in Academia and presented odd thinking.

Somebody once told me that if everyone were a schizoid, abnormal would be those that are considered mentally OK nowadays.

In this way, it can be that, in the end of this book, you think that who has odd thinking is me. In those regards, only God can help us, so have faith.

Perhaps the most important thing is that from thinking that this is funny thinking you end up entering those mental universes, of the people we here refer to, a process that would be called entering the Inner Reality of others by me, and, from that experience, you may learn quite a lot about diversity, multiculturalism, and even complexity. We should all be here to learn, to evolve somehow, so that this is a great opportunity for you to advance a few steps and get closer to what you see in your Inner Reality as your god.

CHAPTER 1
Graham Priest

Graham Priest
ggp@st-andrews.ac.uk

Graham Priest was born in London longer ago than he cares to remember. He studied at Cambridge and the LSE. His first position was at the University of St Andrews. Since only Australia seemed to be prepared to offer him a permanent position, he decamped there, never (at least yet) to return permanently. Cynics will say that he was kept there by being offered chairs in various places. (He is presently Boyce Gibson Professor of Philosophy at the University of Melbourne.) But the truth is that he was seduced by the sun, the wine, and the fact that Australian philosophers are such good fun. It must be said, though, that a few years ago, the University of St Andrews, having forgotten why they kicked him out, offered him an Arche Visiting Professorship. So he returns there for a few months each year looking for sun (some hope), wine, and philosophers who are good fun.

Graham is perhaps best known for his work on paraconsistent logic, and particularly for the heretical view that some contradictions are true. And it is true that he has written a good deal on this, but only because he keeps being drawn back to the subject by the fact that people say such outrageous things about the issue. When he gets the chance, he likes to write about other branches of logic, metaphysics, the history of philosophy, and about quite different things altogether - such as perversion (of a sexual kind, not a logical kind).

When he is not doing philosophy or at the pub, Graham likes to spend time practicing karate-do. He is a 3rd dan in shito-ryu and an Australian National Referee. For a bit of action he likes listening to music, which, he thinks, has been going downhill since Mozart - with the exception of modern jazz.

I met Professor Doctor Graham Priest in the year of 2000 because he was basically the owner of the only course on Nonclassical Logic in the world, like I Googled Nonclassical Logic and I only found his course.

Odd enough was hearing from Graham Priest that I should speak to Da Costa. He said that when we were basically diverging about Paraconsistency. He thought that was a property that could be seen in the ontologies of the actual entities. I thought that was a property that could only be seen exclusively in the theoretical real of things.

Professor Doctor Da Costa was in Brazil, working at UNICAMP, when I travelled to Australia to study with Professor Doctor Priest.

According to the words of Graham Priest to me in that 2000, he was born in Scotland, and he married in England, where he lived for a while. He had at least a daughter from this marriage and she was about my age.

Coincidentally, and for my shock, he actually touched my left shoulder as my father used to touch my knee during a group picture that we took as a consequence of going for a walk together, all of us who were attending the conference at Noosa. We did that during the break from the conference on Logic we there attended. That was 2000.

Despite laughing internally upon hearing about his beliefs, that paraconsistency could be ontological, and assuming that Da Costa had probably called him stupid, since that is a very common attitude in Brazilian science in 2000, I ended up finding a situation where it could be used, and actually wrote a paper about it.

The book I studied in 2000, which was given to me by Priest, mentioned affair a huge number of times, the sigmatoid affair, in Brazil that is considered to be writing of poor quality, it is considered to be a fault, and I also thought it was non-objective language, therefore not something that should be in a book about Science, and I ended up learning that Doctor Professor Patricia Petersen, who Doctor Professor Graham Priest introduced on my first days at the University of Queensland as his partner, like partner for intimacies or girlfriend or biblical spouse, was his affair with time, as he told me he was married in England. He still declared his interest in having the same sort of relationship with me with no shame at my face that year. He had a communist flag over his desk at the University of Queensland during work hours, everyone having to see it and with him teaching inside of his room, which was pretty big. It is just that Australia is a democratic, capitalist Country that signs for human rights when he does that, and, in Brazil, for instance, that is considered treason, so that it is highly illegal, especially because the University of Queensland is a public university. He was still a professor of Philosophy, and, in my mind, Philosophy should be at the university to open our horizons, to increase our range of options, not to reduce them, but any radical regimen, such as communism, rejects freedom by default, so that I did not think he could be working in Philosophy and choosing to be a communist, especially if his main task was make us learn Philosophy. Communism obviously opposes free thinking. Perhaps Marxism would be more acceptable, but not Communism.

I then thought that his Ontological Paraconsistency was more a representation of his Inner Reality than anything else, that is, he was yelling for help by means of that theory: He was highly confused.

Doctor Lea Maria is the greatest psychologist ever alive, and she said that men, when they reach middle age, plenty of them, go through a crisis where they want to still be young, and do not accept, for instance, losing part of their virility. When that happens, according to her, they try to have a woman who is much younger than them to go around with in an attempt to perhaps keep their manhood levels, as for public belief or something.

I personally watched Doctor Professor Patricia Petersen cracking on postgraduate students of my age in our postgraduate laboratory, computers laboratory. I then thought that he was probably senile or going through the middle age crisis, despite looking a lot like middle age was years behind in his human history at that stage.

Thinking like that, I assumed that Ontological Paraconsistency was just confusion because his mind was confused: Being something and not being something at the same time?

We then have a curtain, he would say. Suppose I see blue and say it is blue. You see green and say it is green. It is then blue and green, not blue or green. It is blue and green at the same time, like perhaps the colour is undefined or something.

I said it all depended on our perspective, so that it all depended on angle of sight, position in relation to the object, mental connections or Inner Reality, as I call it now, etc.

I said that the colour was very well defined in the case of the tailor, for instance, since they buy thread of a colour that is represented in a palette that we could perhaps call industrial.

I said that because I had worked at SENAI/CETIQT, which was a mixed economy establishment (maintained by both private sector, industry, and government) that belonged to the Textile Industry and there, whilst teaching Mathematics, I had gotten interested in visiting the tailors and studying a bit of what they do.

Every colour that we see in a fabric is defined in an industrial palette, so that each one of them has a specific and unique name, and there is no possible confusion. Yes, our eyes can get confused and classify a turquoise blue as a deep blue, for instance, or even as a green of some sort, but the industrial palette would define the colour we see in the fabric as colour number 322, say, a unique and distinct colour.

For me, therefore, his thinking was simply confusing and originated in a person who obviously had never bothered understanding the simplest thing about what he seems to think he masters.

For me, Science was Science, and, if we solve a problem in a way that nobody has doubts about it, then it is solved.

The colour of a piece of fabric was in that category: If someone manufactured it, we knew exactly the colour it had.

It is possible that I, Doctor Professor Marcia Pinheiro, had no notion about the particular number involved, but I at least knew that such a number existed, and that made Priest's statements impossible to be believed from first second.

Other examples existed, however.

One of them, which he liked quite a lot, involved a painting of the type illusion: There were stairs that nobody could tell for sure went up or down.

I then said that it was all a matter of perspective, like it mainly depended on what we had in mind, but other factors were also involved, so say position of our bodies during the observation, eye health, etc.

I said that when a person said that they went up they had a certain referential in mind, say that when you know something goes up the foot of the person going up through that something is to the right, and when they said they went down they had another, so say the head of the person walking over the object would be in alignment with a certain object.

In this case, since the referential was not the same, it couldn't be the case that things were and weren't at the same time: Things were at a time and things were not at a different time in our body clock, let's say.

Priest seemed not to have argument against this one, so that, to me, it looked as if he had given up on publishing about the topic because of our conversations in that 2000.

Doctor Lea Maria would finish with his conversation, when he was talking about the curtain, but she is a psychologist, by saying that he was probably daltonic, I reckon.

After I conversed with Trevor Skinner, in the end of 2001, and mentioned our exchange of scientific tokens, and I then also mentioned Koji Tanaka and how he could write about Priest's thinking, Ontological Paraconsistency, Doctor Professor Tanaka actually published a paper talking about that.

I met Dr. Koji Tanaka at the University of Queensland. He was a postdoctorate student for Priest in 2000.
In my Inner Reality, Priest invited Tanaka to study with him because I was not fulfilling his expectations, which were probably that I continued his work, that I did what he did. I think that, in his view, his work was creating new logical systems, since that is what Koji ended up doing, as for what I saw of his work in that 2000.

I was not necessarily rejecting doing that. It is just that I had basic disagreements and would like to work on the foundations, as I finally did, despite what I endure, and only because of what I endure it was as late as you see in those papers (Entailment, Completeness, and Ontological Paraconsistency).

I asked Priest to enter professional research, to have his aid to do that. I did not ask Priest to work on what he worked. It is all about expectations that people create inside of themselves, I reckon.

I said I wanted to learn about Nonclassical Logic, since, first of all, I only found his course, like in the entire world, as therefore what they later on would invent was a Googlewack. This sigmatoid, Googlewack, would actually also have originated in my conversations with that demoniac man, Trevor Skinner: I did report to him on how I ended up in Australia (several people asked me this question, and I think I now also know why, I would say very unfortunately. Perhaps the question that they asked would equate, how a person like you ended up in Australia, I now think).

I learned about this new sigmatoid, Googlewack, because of a comedy show. One of the things I most loved in Australia was the comedy shows. Not only I thought I wanted to present one, as this lady, who much impressed me, Sarah Kendall, did, and I watched her first show of that type I think, but I thought it was a fantastic way to have fun, something that we really did not have down in America, at least in Brazil.

I told Trevor that when I tried Nonclassical Logic, and, as I keep on saying, that was for reading Simon Singh's book, Last Fermat's Theorem, together with the sigmatoid course, I found only the University of Queensland, like in the entire world, so that Dr. Google, if you so wish, had answered, University of Queensland has it, when I used him as an oracle, basically.

When I opened the site Dr. Google so carefully had put in front of me, as its Googlewack, I saw contact details that corresponded to Priest's, and that is how I ended up writing to him the first time.

I was really excited with all because Zelia Cardoso de Mello had performed the miracle of equating the Brazilian currency with the Australian dollar. Her exotic formation in Economy and her intellectual beliefs made her form the most successful Economy team Brazil ever had.

With that, it was actually affordable to me.

Dr. Priest however was facing his own difficulties. I would learn, by the first semester of my course at the University of Queensland, establishment I much liked and respected until the end of the first semester, that Priest was being asked to finish with precisely that course, the one on Nonclassical Logic. The Chancellor, John Hay, was asking him to do that.

When he saw me interested in his course, writing to him from Brazil, and saying that that was basically a Googlewack, he immediately thought of the use I could have in his life: Being me an international, someone from America, and from a popular Country at the University of Queensland (I myself met at least 40 Brazilians who were there that year doing postgraduate courses, this in an event promoted by the own university that Priest told me to attend), he thought that he could use me to keep the course happening at the university.

This is what we could call inner assumptions, things that belonged exclusively to his Inner Reality.

I came totally unaware of that, and, from my end, I was paying a hell lot of money for that course, in fact the only real estate property I had, so that I really had to make great use of that.

He was all smiles and welcomes from beginning to end of our exchanges in terms of me coming to Australia and doing the course with him, and I was thinking that was only good manners, good customer service, and what he was supposed to do being the manager of the course.

I let it clear that my intentions were those, and therefore had nothing to do with praising him as a researcher or his course at the university or anything else: My intents had to do with my own development and career. I had been told, in an informal conversation with University of Queensland representatives in a postgraduate fair in Rio, that their courses were wonderful because basically I could shape them in any way I wanted, and that was my worst problem in Brazil in terms of postgraduate courses: I wanted to simply get into research, do what I wanted to do in research, but the educational structure of the Country was such that they copied the Americans, and therefore had absolutely inflexible system, not made for what we want, but for what they want instead, like saving resources, enlarging use, etc.

I actually knew nothing about his work as a researcher before I came. I had read Simon's book, got incredibly interested in what was new in Mathematics, what sounded like revolution, and I then came.

I now believe that he thought I knew about his research before I came. I think I am sure he assumed that I would not deny him the favour of convincing the Chancellor that his course was a good thing for the University of Queensland, and, worse, he thought I could be very happy in finding in him a husband or a sexual partner.

Reality is well another, but I do think his Inner Reality had the entire paragraph settled.

When the second semester came and my passion became The Sorites, and again not Professor Doctor Domenic Hyde, but The Sorites, his disappointment was extreme.

I think he also expected, as said before, that I was going to follow his research ways. That was probably his most normal expectation, most acceptable, I think.

Because I was worried about foundations, refused to pay compliments to his course before the chancellor in the first semester, and instead asked to see how the second would go, and, in special, the end of the course, and was also not interested in doing what he was doing, first of all worried about the utility of things, he stopped thinking he should support me in Logic, I now think.

Yet, he also had work on The Sorites, I later on would find out also by means of writing. In 2000, he told me he had also produced work on The Sorites, so that that should not be an exclusive thing for Doctor Hyde, but, in 2000, he never presented any piece of his writing about the problem.

I got extremely injured in that second semester by his choices, and I saw nothing in that but the tag harassment.

It was in the second semester that Koji appeared. Koji is from Japan and I had some huge sympathy for the Japanese at that stage of my human incarnation for a few good reasons: my grandmother had a home seller from Japan, a man who she always complimented, who came to her home to basically dress her. I was told by God to join Seicho-no-ie, a mysterious process I describe in one of my books on religion, and that is a Japanese religion. I read their philosophy, listened to their teachings, and thought they were correct in all I saw and heard. And I had read a book called Shogun, which I found in one of my visits to bookshops in Brazil. This book made me travel to Japan and live there, with those people. I basically felt as if I were a Japanese woman and dressed shoes with which I could not walk. I also felt as if I were the warrior involved. The book also taught me some Japanese on top, made me get interested in their language. That is when I learned the value of writing things in that way, what also made me write Terra Australis (about Brazil and Australia instead. Some Asian culture, small token). The art of doing things well is something my mother, Dr. Lea Maria, was always making me pay attention to. Shogun was perfect writing for me: entertaining, knowledgeable, cultural, exciting, adventurous, mysterious, etc. I actually ended up giving either that book, like a new one, or the one that came after that, to a manager I really liked at Itau, the bank I inherited from my father, General Braganca. Her name was Lilian, and, upon thinking about what could please a woman who was like me, that is, a professional woman, with ambitions that go well beyond having romance, a woman of a certain level of culture, which goes well beyond the level of the vast majority of the women in Brazil, and a woman with class, which also goes well beyond what is normally seen in Brazil, it occurred to me that a book was good thinking, like I would not spend much money, which is what I simply had to do, and it would be really meaningful. I totally think that the feelings we imprint in our hearts and souls because of the words of others that we ourselves read are way more unforgettable than any other thing we can experience which is not intimacy. I think that the more we can travel through the writings of someone else, that is, through their hands, the more we can inhabit the possible worlds of Priest, actually.

I had noticed that Lilian was the second best there (the best was the manager that served my father, a man who made me go with Lilian, perhaps because he felt sexually attracted to me or thought he could, and he was a good family man, perhaps because of his wife or something), but she received gifts from all the clients that she liked, like I only saw her smiling and happy when they arrived with a gift to the side. I was five stars because I also inherited the stars from my father, but at least sometimes we had to wait for her, even though we could go straight to the manager, skipping all cues.

On those occasions, I would observe, since I had nothing else to do, basically, or nothing else that I could do. That is when I thought of gifting her myself. And I noticed that it did work in the same way, so that she was my person: If I know how to make them happy, and that is something I can do, then that is good enough for me, like what I really hate is mental, people of the sort there is nothing you can do to please, like all you do is worthless and returns no happiness, no gratitude, etc.

I then liked her: She was simple. Now I was not so upset with the exchange (my father's preferred manager for her) anymore. I liked his cross, to the back of him when he was sitting at his table, and that connected to my Catholic background, schools, etc., but now she was a suitable replacement, finally.

Oh, well, this is to say I liked Koji because I really sympathized with the Japanese that far. One of the things that most impressed me when I first arrived at the Seicho-no-ie temple, in Catete, Rio, was that the Asian man who was lecturing was a person of God: He was saying that it is our ancestors that bring us to Seicho-no-ie when I entered the temple's door. Had we synchronized from training, it would not be that perfect, quite sincerely. His face was entirely turned to my face as I entered, and things were in such a way that it was exactly as if he were talking only to me.

I was hours away from the experience that led me to Seicho: I had done the three-day fasting technique from the Catholic bible, technique you see in The New Rosary (Amazon.com), and I had woken up in the middle of a circle of hooded people dressing brown. I saw no face, feet or hands. A voice came to my right ear and whispered Seicho-no-ie. The three-day fasting technique involves mentioning our ancestors and then making a wish. I asked God to tell me the religion I should follow, the religion that most agreed with His teachings, but demanded no effort, a religion that basically did not offend Him. I was sick of trying churches and ending up really upset for just wasting my time. I did not like the Catholic Church anymore and for long. I basically slept during the masses when I took grandma there on Sundays. I found it hard to believe the priest ever cared about grandma, so that his teachings were not followed by himself.

I had never heard that expression or word or sigmatoid in my life, but, as the Australians say, I decided to give it a crack: Whoohoo, what a surprise! The Yellow Pages had it. It was indeed a religion!

I get there, the preacher is saying that sentence, so that it all looked like it had to be God.

On that same day I heard the same blessed preacher saying that we should thank the chair for being able to sit on it, just because basically what was there was the love of the person who built it. I never heard something more inspired coming from the mouth of a man that far (usually all those tokens would come from my mother's mouth). I was marvelled.

With that, I started thinking about all that had gotten to me from such a distant culture and people: some killed themselves at very young age for underperforming at school, and that meant getting anything that were not ten out of ten (I get incredibly upset if I don't get those myself). They created Macrobiotics, which was a diet my father had adopted. My father actually lectured me a little on that. They possessed an entire suburb in Sao Paulo and Brazil never bothered not even obliging them to speak Portuguese, the national language (I went there personally. They did not speak Portuguese at all, an entire suburb in one of the most important places in Brazil, if not the most important one). And there were still tons of other small things, such as the home seller and how happy grandma was with his service provision and basically love, using Seicho Talk, let's say.

The details: The Japanese praised the details, just like my mum. Shogun taught me about rites. They had respect in their culture, respect for several things.

What I most found annoying in my birth family and in Brazil, especially Rio, was disrespect: Those savages seemed to have complete disrespect for work, for human effort of all sorts, for love, especially of real nature, etc.
My mother, my grandmother, Jayme, my father, myself, Murillo, we all seemed to be targets of their disrespect all the time: Regardless of the effort we put into building something, pleasing them, and others, they seemed to always have immense disrespect for us all.

My mother asked Lea Pinheiro to simply clean her own bedroom periodically and leave its door open, so that she could circulate in the apartment she herself paid and cared for. I saw that as minimum effort, and definitely always did it. Lea absolutely never bothered, like not even cleaning, not even once.

At a certain stage, her bedroom had no place for us to put the foot in. She forced me to enter it with her, as she frequently did, perhaps for fearing the monsters that could appear from inside of her own mess, quite sincerely, and there were even menstrual pads with blood on them all over the top of whatever hell she had on the floor, and the mess was about waist high.

She kept on saying that I and mum were insane, but I thought to myself, very early in life, that if someone was insane, that was obviously her.

That was minimum effort. The day she got basically evicted, I said thanks God and mum for the next 24 hours in a row.

She complied with nothing.

Having had contact with people like Lea Pinheiro, Agnella and Joao Terra, Hermolga, and others my entire life in Brazil before coming to Australia, I really loved the Japanese in all I knew about them: Just the fact that they could respect the effort of others, and they cared precisely about the details, so say how long I myself spent basically building this book, meant the world to me.

Just the fact that they could observe human subjects for real, and put maximum effort in whatever they did was a big deal: I watched their writing and I remember thinking that it was very unlikely that I could ever do that.

The art in their hands, the care with the strokes, all so perfect, the art in their fabrics, and their concerns seemed to be just the right thing. I remember having been gifted a dress that would basically last forever without any effort put into that.

That was Japanese silk that my grandmother had found in the Japanese's man bag: He always came around with a paper bag, usually of brownish or greyish colour. I dream of a world of that fabric since then.

I basically hate washing, ironing, and things like that, and I do follow The Bible, so that I also hate forcing others to do that for me, regardless of how much I would be paying them to do that.

They seemed priceless to human kind.

Ironically enough, it was my ideas and Science that have been used to destroy their Country and people: I designed all for Rio de Janeiro, and told Trevor in a generic manner what could be done to simulate a natural disaster and finish with Rio before they finished also with the First World and the rest of my existence.

Very unfortunately, since then, those tokens have inspired a tsunami that took 200,000 lives from Indonesia and another disaster, of similar nature, that took thousands of Japanese lives.

It still seems to me that it is quite possible that Julia Gillard conversed with the Japanese leaders and they told her to rescue my rights and give me justice before the first incident of that nature hit their Country, since she seems to have travelled to there shortly before it, and the American Seicho leader had become aware of all I endured, that was a woman, contrary to Seicho beliefs (it should be a couple, but all was destroyed by what I call now The Trevors), and I did ask her to ask the Japapense leaders to intervene because, first of all, she seemed not to know English.

I had asked Trevor to bring Seicho to us somehow, perhaps to the United States, but I had also asked people, such as Kate Morioka, a girl I thought was a true friend of mine in that end of 2001, Australian-Japanese, to contact them.

I recently sent my own English version of one of their songs to them (blue book in Brazil, Marcha da Missao). It is all really positive, quite sincerely.

Anyway, it was in this spirit that I mentioned Tanaka to Trevor in that 2001, and, amidst that amount of confusion, in which I ended up wanting to even prosecute Priest, given the horrible way the support staff from VUT, all native Australian women, had dealt with my issues, quite criminal: I then hoped that Tanaka could help me publish about Priest's theories and writings, in a way that Priest would not be upset and I would not be injured.

If there is one thing that I hate in life is wasting resources, and I saw a lot of wealth in the time I had spent with Priest, one of the most remarkable thinkers of the 21st century. I just wanted to make sure I could optimize my own efforts there.

I feared that he would feel offended with my writing, and would then not like me anymore (I did not know yet the depth of the results of the actions of the support staff from VUT, like he was already offended well beyond belief), or that he would deny having told me things in what came to Ontological Paraconsistency, since I found nothing that he had published about the topic.

I then suggested that we used Koji, who seemed to be more liked by him, and then got something published in what regarded Ontological Paraconsistency.
I also had the intuition that Da Costa, from UNICAMP, had offended Priest by perhaps saying idiota to a fellow in Brazil as Priest defended his Ontological Paraconsistency in front of him. Perhaps unaware of the English language, and how much this sigmatoid sounded like what its translation is in the English language, Da Costa would have answered to a fellow's query, and, to be short, but still close to Priest, he would have released the bomb: Idiota! In this way, and believing that Priest did sound stupid on at least two occasions, which were when he was explaining Ontological Paraconsistency and defending that The Monty Hall Problem presented a problem for the acceptance of the foundations of Combinatorics, I suggested (to Trevor, 2001) that the Brazilians built a robot using Paraconsistent Logic, which would then be a robot that would simply act upon receiving contradictory information from the environment. I suggested that humans entered the instruction to be followed. The Brazilians actually did that, and I mentioned that in my work. They did that during what I am calling my martyrdom, which is a period that starts in the end of 2001 and is still going. I knew that the Brazilians would not reject doing that. That was to exemplify their own theory (Da Costa's), which would then be the Non-ontological Paraconsistency, as I explain in my work, which is also in this book.

I was totally convinced that I loved Australia and all these guys did here in research, since it was all at my reach, very different from Brazilian Science, in all I got interested in, first of all. I just wanted us to be happy people in all, as always, as happy as common sense lets us be, as my good mother used to say, and I thought that I am transparent in all I think, since my mother kept on telling me that, like she seemed to think that I never needed to say words to communicate, since my face was really expressive. I feared that to death in what came to dealings with academics in Australia, since if I thought the wrong thing, they could finish with my career.

Priest, at a certain height, told me to speak to Da Costa. I then thought, since that immediately followed me expressing my opinion about his Ontological Paraconsistency, discussion that Patricia Petersen somehow got to know about and understand for odd reasons (perhaps not in full), that he OBVIOUSLY, because of my mum's opinion, Dr. Lea Maria, saw, at my face, that I was thinking IDIOT.

That was then my strongest theory when I spoke to Trevor: Sincerely, those guys in Brazil would call people idiots quite easily. I definitely think they would call all our Australian discussions stupidity, quite sincerely, but I absolutely love all this, for what I worry about is precisely these things, first of all.

I ended up losing my precious life, turn, careers, and perfect head and body for atrocity, but the intentions were completely the opposite when I spoke to Trevor, believe it or not.

Sometimes things just don't go in the way we intended, but we also don't expect to get what I got not even for a second, is it not?
I don't get my post since at most 2005 in Australia, for instance. I never got not even one letter from Trafford. Yet, as far as I know, they should be sending them every year. At most 5% of the items destined to me ever get to me since then, believe it or not. This is the smallest thing I endure. If you hear about the rest, you may spend the rest of your life crying.

Anyway, Koji's paper is on the next page.

Three Schools of Paraconsistency

Koji.Tanaka@mq.edu.au

Received by Greg Restall
Published July 1, 2003

Abstract: A logic is said to be paraconsistent if it does not allow everything to follow from contradictory premises. There are several approaches to paraconsistency. This paper is concerned with several philosophical posi-tions on paraconsistency. In particular, it concerns three 'schools' of para-consistency: Australian, Belgian and Brazilian. The Belgian and Brazilian schools have raised some objections to the dialetheism of the Australian school. I argue that the Australian school of paraconsistency need not be closed down on the basis of the Belgian and Brazilian schools' objections. In the appendix of the paper, I also argue that the Brazilian school's view of logic is not coherent.

But though logic has come a long way very recently, it has a longer way to go, both in *whom* it involves and *what* it investigates. There are, for in-stance, virtually no black researchers, and exceedingly few women are en-gaged; and for all the proclaimed rationality of modern humans and their institutions, logic touches comparatively little human practice. Differ-ently, there remain *many* notions of considerable logical import, *some* of historical significance, of which we lack decent accounts or, sometimes, a clear appreciation. To the satisfactory elucidation of these, sociative logics can make essential contributions.

Sylvan (1989) p. 133.

1   INTRODUCTION

A logic is said to be paraconsistent if it does not allow everything to fol-low from contradictory premises: it is not the case that for any α and β,

---

*I would like to thank Graham Priest for making some of the ideas contained in the paper clear and for his comments on drafts of the paper. I would also like to thank Diderik Batens, Otávio Bueno and Newton da Costa for their comments on a draft of the paper. A draft of the paper was presented at the 1998 Australasian Association for Logic Conference held at Macquarie University in Sydney. Many thanks go to the members of the audience.

∴ β (*ex contradictione quodlibet*).   There are several a
to achieve this end, as is surveyed by Priest and Routley (1989a) and Priest and Tanaka (1996). This paper surveys several philosophical positions on paraconsistency. In particular, the paper concerns three 'schools' of paraconsistency: Australian, Belgian and Brazilian.[1] Arguably the most radical school, the Aus-tralian school of paraconsistency, led by Priest and Sylvan (né Routley), claims that there are some true contradictions and that *the* logic is paraconsistent. (Sylvan did however advocate logical pluralism in a posthumously published work, Sylvan (1997).) The Belgian school, led by Batens, and the Brazilian school, led by da Costa, argue against the Australian school. They question the existence of true contradictions. More importantly, they not only reject the idea that *the* logic is paraconsistent but also deny that there is a uniquely correct logic. I argue that their objections are based on misinterpretation of the claim of the Australian school and/or are unsuccessful. I conclude then that the Australian school need not be closed down on the basis of the Belgian and Brazilian schools' objections. Moreover, in the appendix of this paper, I argue that the Brazilian school's view of logic is not coherent.[2]

2. SCHOOLS OF PARACONSISTENCY

If we put aside the 'forerunners', the first paraconsistent logic was developed by a Polish logician Jaskowski′ (1948). His approach was to not allow premises to be adjoined: $\{α, ¬α\}$ 6|= $α ∧ ¬α$. This non-adjunctive approach has remained in the Polish school of paraconsistency.[3] The non-adjunctive approach has also been advanced by the Canadian school of paraconsistency.[4]

After its inception, the development of paraconsistent logics has been car-

---

[1]As was pointed out by a number of my colleagues, referring to schools of paraconsistency in terms of geography is not appropriate. For the ideas held by paraconsistent logicians are not bound by geographical borders. However, the schools described herein are 'typified' by practitioners within each respective country, and I therefore take the labels to be illuminating and at least partially descriptive.

[2]Parts of the paper can be seen to deal with the issue of logical monism and logical pluralism. However, I do not touch upon that issue for two reasons. Firstly, the contemporary debates between logical monists and logical pluralists are recent phenomena. The rigorous debates between them started with an explicit formulation of logical pluralism and a rejection of logical monism by the Australian logicians Beall and Restall (2000). This paper was written well before the publication and even the explicit formulation of their position, and hence I was unable to benefit from the debates between logical monism and logical pluralism. Secondly, the issue that I am concerned with in this paper is not that of the debate between logical monists and logical pluralists in general. It is rather the objections to the Australians that have been raised by the Belgians and the Brazilians. Dealing with the issue of logical monism vs. logical pluralism takes us beyond the scope of the paper.

[3]See Perzanowski (1997) for the Polish school of paraconsistency.

[4]For example, Jennings and Schotch (1981) and Schotch and Jennings (1980). The non-adjunctive approach was also taken up by the Americans, Rescher and Manor (1970) and Rescher and Brandom (1979).

ried out in many different places.[5] Arguably the most radical approach has been taken up by the Australian school of paraconsistency. The major roles have been played, among others, by Priest and Sylvan who claim that there are true contradictions. They argue that because of the existence of true contra-dictions, logic must be paraconsistent.[6] This dialetheic approach has raised eyebrows among many people including many paraconsistent logicians. The Belgian school, led by Batens, has objected to the claim that logic must be paraconsistent. The Brazilian school, led by da Costa, has questioned the claim that there are true contradictions and also rejected the idea that logic must be paraconsistent.

3. LOGIC AND LOGICAL SYSTEMS

Before examining the Belgian and Brazilian schools' objections to the Australian school, an important issue in the philosophy of logic needs to be ad-dressed. That is the distinction between and logical systems.

Now, I would be, quite correctly, accused of being absurd if I were to defend the view that the theory of dynamics *is itself* moving entities.[7] The theory of dynamics is an explanation and description of moving entities. Explanation and description are *not* themselves moving entities. As Restall (1994) puts it:

The general theory of relativity may describe the Way the World Is in a clear and perspicuous way, it may fit the facts, or be ide-ally useful, or maximally coherent, or whatever – but it isn't to be identified with what it is intended to describe. (p. 11)

There is no need to elaborate on this issue any further.

However, it is not clear at all whether the point is widely taken when the topic is logic. It is in fact a common confusion in logic not to distinguish a theory from entities which are described by it, where the theory in question is a logical system which has mathematical properties, and the entity is .[8] In advancing a pragmatist conception of logic, Haack (1996) describes a logic (i.e., a logical system) as a theory: logic [i.e., a logical system] is a theory, a theory on a par, except for its extreme generality, with other, 'scientific' theories . . . (p. 26)

Haack does not mention in her discussion. Yet, just as in the theory of dynamics, a logical system must not be confused with . For a logical system is a theory of and hence it is not itself .

---

[5]See Arruda (1989) and Priest and Routley (1989b).

[6]Not all of the Australian paraconsistent logicians hold this view. In this paper, I am only concerned with the approach taken by Priest and Sylvan.

[7]The example is taken from Priest (1987), as will be clear later on.

[8]What the nature of is, is an interesting question. However, I leave the entire ques-tion for another occasion, except that can be thought of as a truth-maker of a logical system.

The story is not as simple as it is presented above. In the case of science, there is no dispute that there are some entities that a theory describes. Even if one is not a scientific realist, no one seriously argues that the theory of dynam-ics is itself moving entities. Now in the case of logic, it is not uncommon to reject the existence of . For, unlike moving entities, is not any-thing material. The logical instrumentalist argues that there is nothing that a logical system corresponds to. For them, the study of logic is no more than that of internal properties of logical systems such as the systems' algebraic proper-ties. None the less, the logical instrumentalist does not claim that a logical system *is* . They simply do not accept the existence of . If one is a realist about logic, whether a monist or a pluralist, and so believes in or s, then the difference between and a logical system must be apparent. That a logical system is not itself is now well established.

4. ...AND THE AUSTRALIAN SCHOOL OF PARACONSISTENCY

The Australian school of paraconsistency takes seriously the distinction be-tween and logical systems and argues against classical logic. They argue that classical logic is a theory. Since it is a theory, the idea that classical logic could not even be questioned must be rejected. As the history of science tells us, a theory may be shown to be false at a later time regardless how well the theory is entrenched. Priest (1987) writes:

No one needs to be told that one needs to distinguish between our theory of dynamics and moving bodies themselves. One is an attempt to provide a correct theoretical explanation for, and de-scription of, the other, and to confuse the two is absurd. Yet a similar confusion is common in logic. The fact that we use the same word, 'logic', for both is but an effect (rather than a cause) of this. But just as with dynamics, so with logic, one needs to distinguish between reasoning, or better, the structure of norms that govern valid/good reasoning, which is the object of study, and our logical theory, which tries to give a theoretical account of this phenomenon. The theoretical principles we do actually accept are not God-given and fixed for all time. Indeed, reasoning is a complex and delicate human activity, and it is unlikely that any theory we produce, at least for the present, and maybe for ever, cannot be im-proved. The norms themselves may also change. There may well occur a dialectical interaction, characteristic of the social sciences, between the object of the theory and the theory itself. None the less, the distinction between a science and its object remains; and once this gap is opened, it suffices for the fallibility of any theory. (pp. 257–258)

Restall (1994) uses exactly this thesis of Priest to meet Quine's famous objec-tion that changing the logic is changing the subject.[9] Based on the distinction between and logical systems, then, the Australian school of paraconsis-tency argues that classical logic must be rejected as a false theory.

The problem that the Australian school points out is that classical logic fails to capture properly. In order for a logical system to do so, it has to encapsulate truth-preservation in *all* situations, as Priest (1999a) argues. What situations need to be considered and how large the realm of is, are the questions that invite hot disputes, as Haack (1978) demonstrates. The Aus-tralian school of paraconsistency argues that all situations, whether consistent or not, need to be considered.[10] For, Priest (1987) and Routley (1980) argue in their discussions on semantic paradoxes and set theoretic paradoxes, the realm of includes inconsistent situations. Yet classical logic fails to capture truth-preservation in inconsistent situations.

However, Priest (1989) argues that the use of classical logic is acceptable in some situation *if* there are good reasons to assume that the situation is con-sistent. For him, consistent situations have a peculiar (al) structure and classical logic accommodates this peculiar structure. This does not mean that paraconsistent logics do not capture the structure. Classical logic can be seen as a special case of paraconsistent logics: classical logic can be 'recaptured' from a paraconsistent perspective.[11] The recapture demonstrates that the pe-culiar structure is well within the reach of paraconsistent logics. None the less, classical logic fails in inconsistent situations. Hence Priest (and Sylvan) argues that the "universality of classical logic must be rejected".[12]

Since logic is about truth-preservation in all situations, therefore, paracon-sistent logic should be accepted, at least tentatively, as a correct logical theory instead of classical logic. For some evidence shows that it is paraconsistent logic that accommodates all situations and hence captures properly. Thus Priest and Sylvan argue for the universal validity of paraconsistent log-ics.[13] Whether this Australian school's claim is ultimately defensible or not, I do not pursue in this paper. Instead, I will consider some objections to the Australian school. Those objections have been put forward by the Belgian and Brazilian schools of paraconsistency.

---

[9]Quine's objection is found in Quine (1970).

[10]A paraconsistent logician may be concerned with not only inconsistent situations but also incomplete situations. As issues surrounding contradictions are the subjects of this paper, in-complete situations need not concern us here.

[11] See Priest (2002) for the classical recapture. Mortensen (1995) argues that classical mathe-matics is a special case of paraconsistent mathematics.

[12] Priest (1987) p. 257.

[13] See especially the appendix of Routley (1980) titled 'Ultralogic as Universal?' where Ul-tralogic means one particular paraconsistent logic. See also Mortensen (1983) for his defense of the non-validity of *Disjunctive Syllogism* which supports the claim made by Priest and Sylvan.

5. ...AND THE BRAZILIAN SCHOOL OF PARACONSISTENCY

The clearest statements of the Brazilian school's objections to the Australian school are found in da Costa and Bueno (1996). They argue that the views of the Australian school of paraconsistency are fallacious. Their argument is that paraconsistent logic is not the true logic on the basis that any logic may be false. However, their objections to the Australian school are based on a misconception of the philosophy of the Australian school.

In comparing classical logicians and non-classical logicians, i.e., Australian paraconsistent logicians, da Costa and Bueno reject the idea that paraconsistent logic is the true logic. They write:

Instead of claiming, with the older «radical» proposals, that *classical* logic is a tool to apprehend the most general structure of the world (it is supposed to be true after all!), their new «radical» version [i.e., the Australian school of paraconsistency] claims the same as far as certain *non-classical* logics are concerned. One wonders, in such a case, about the meaning of learning from experience — that is, from the recent history of logic. Indeed, what is the import, the relevance of this history, with the changes and moves that it has yielded, to our philosophical understanding of logic? Given these circumstances, how not to be fallibilistic after all? (pp. 53-4)

This complaint seems misguided, at face value. As we saw previously in the writing of Priest, the Australian school advances a fallibilistic view that classical logic is a false theory.

One may argue, however, that the Australian school is *not* fallibilist about paraconsistent logics. In this sense, the Australian school is the same as the school of classical logicians. Yet this is to misunderstand the claim made by the Australian school. What they are advancing is a fallibilistic view of *any* theory. Their view is not that of the dogmatist. They do not dogmatically hold that paraconsistent logics are true theories. Their fallibilistic view of logical theory includes the fallibility of paraconsistent logics. Based on their analysis of several issues, the Australian school claims that it is paraconsistent logics, not classical logic, that should be embraced. For, given the currently available data, paraconsistent logics solve the problems that they consider, such as se-mantic paradoxes and set theoretic paradoxes. None the less, they admit that they could be wrong, as paraconsistent logics are theories.

The mistake that the Brazilian school makes is not only that they take the Australian school to be dogmatic but also that they take the Australian school to be claiming that logic is "stable".[14] Kant thought that logic was a completed science. He held the view that logic "contains merely the form of thought" and that the laws of logic are "the conditions of the use of the understanding in

---

[14]da Costa and Bueno (1996) *passim.*

general".[15] Based on these views, Kant argued that logic could not be altered. There are two things that have to be said about the Brazilian school's claim.

Firstly, a logical system is a theory for the Australian school. Since it is a theory, a logical system could be altered in order to accommodate new evidence. For example, one may decide to eliminate some classically accepted rules of inference. This is how substructual logics, some of which are paraconsistent, were developed.[16] Secondly, the Australian school does not argue that is stable. We have seen Priest arguing that the norms that govern

reasoning may change. There is no argument to the effect that does not change over time.

Thus, objections put forward by the Brazilian school of paraconsistency to the Australian school are based on a mis-conception. Whether or not the philosophy of the Australian school, properly understood, is defensible, is another matter. None the less, da Costa and Bueno's objections can now be put aside.

6. …AND THE BELGIAN SCHOOL OF PARACONSISTENCY

The Belgian school of paraconsistency is concerned with the Australian school's claim that is unique. They argue that there are several s each of which has its domain.[17] Consequently, they argue that there are several logical systems that are correct. In objecting to the Australian school, the Belgian school proposes contextualism. Batens (1990) writes:

I think there is another alternative [to the Australian approach] for which there are good independent arguments, viz., 'contextualism'. The idea is that we do not depend on a fixed global system, which should be justified once and for all, but that we set up a specific context (involving meanings, relevant data, methodologi-cal instructions, etc.) whenever we meet a problem. (p. 226)

That there is an alternative is not sufficient to reject the Australian approach. In fact, contextualism is not in general incompatible with the Australian approach. For an analogy, consider the case of geometry. It is often said that different geometries are

appropriate for different contexts. For example, when we build a house, it is appropriate to use Euclidean geometry. But when we do surveying, it is appropriate to use spherical geometry. And when we do astronomy, it is appropriate to use Riemannian geometry. However, it may be the case that is unique and that the above three geometries capture it in only some situations. Some factors may safely be ignored when a geometry is applied to a different context. The same may be true of logic. As we saw earlier, the Australian school argues that paraconsistent logics and classical logic

---

[15]   Kant quoted in Haack (1996) p. 27 and p. 28 respectively.
[16]For an introduction to substructual logics, see Restall (2000).
[17]The Brazilian school seems to hold a similar view. But their view is not as clear as that of the Belgian school.

collapse into each other in a consistent (and complete) situation in the sense that both paraconsistent and classical logic capture the consistent (and com-plete) part of . The acceptance of the use of classical logic in consistent (and complete) situations does not interfere with being paraconsistent as a whole. Hence the Belgian approach is not necessarily incompatible with the Australian approach, as far as logical systems are concerned. But based on contextualism, Batens puts forward another objection.

Batens takes to heart the Australian school's claim that the correct logic is the one which captures the logical aspects of all situations. Despite the Australian school's claim, he argues that paraconsistent logics are too weak to be applied to consistent situations. Batens (1989) writes:

If applied to consistent sets of premises, paraconsistent logics lead to proofs that are considerably poorer. (p. 190)

Specifically, he argues that there is some 'context' in which paraconsistency fails.[18] Batens (1990) writes:

The metalinguistic description of classical negation is beyond the reach of paraconsistency, and so is classical triviality. (p. 227)

In order to analyse Batens' objection, we need to be clear about what he means by a context. He defines it in Batens (1985) as follows:[19]

What I mean by 'a context' is precisely a communication situa-tion. A context is characterized by (i) a set of participants, (ii) the problem that one tries to solve, (iii) the set of statements that are regarded as certain and in this sense define the set of possible an-swers to the problem, (iv) the set of aspects that are considered relevant to the problem, and (v) the set of methodological do's and don'ts that are judged appropriate with respect to the problem. (p. 334)

Based on this definition, Batens argues that each context has its own and a in one context may not be the same as a of another con-text.

Now the context that is beyond the reach of paraconsistency, that Batens has in mind, seems to be the one in which 'negation' is expressed in terms of 'classical negation', which we denote by $\sim$.[20] Let's call this context c. Then Batens' objection is that the negation operator of, for example, Priest's logic (1979), which we denote by $\neg$, is too weak for context c. Thus the of c is not paraconsistent, at least not .

---

[18]   Parsons (1990) raises a similar point.

[19]   Batens (1985) is mainly concerned with the philosophy of science. But Batens (1992) claims that his contextualism is substantiated in logic, and Batens (1985) presents a contextual philoso-phy of logic towards the end of the paper.

[20]   In the literature on relevant logics, $\sim$ is called Boolean negation.

The Batens objection may appear intuitive, since ~ and ¬ are *prima facie* incompatible with each other.[21] Yet its justification is hard to come by. Since it is a classical negation, ~ is defined as follows: ~α is true iff α is not true. This definition satisfies the law of non-contradiction, the law of excluded middle, and so on. On the other hand, ¬ is defined as follows: ¬α is true iff α is false. If we assume that a formula cannot be neither true nor false, this definition also satisfies the law of non-contradiction, the law of excluded middle, and many other laws of classical logic, as Priest (1999b) demonstrates. Moreover, if inconsistency (and incompleteness) is rejected,[22] this definition will be equivalent to that of ~. For, in this case, if α is not true then it is false, and vice versa. Hence ~ and ¬ behave in the same way. In particular, ~ and ¬ are identical in context c. Thus, if ~ captures the negation fragment of of c, ¬ does the same.

Still, Batens argues that ¬ is weaker than ~ in context c. He claims that ¬ cannot be used to express the exclusion of α by ¬α and vice versa. Batens (1990) argues that if negation is expressed by ¬,

[a sentence asserted] does not rule out the sentence that is negated and is intended

not to rule this out. (p. 223)

Indeed, the fact that α and ¬α both may be true seems to show that expressing exclusion is beyond the reach of ¬.

But how is the exclusion expressed in classical logic? The definition of ~ does not do the job. For if α is both true and not true then both α and ~ α are true. According to Batens (1990), the exclusion is expressed in terms of , and is the only way to express it. For, by adopting

Someone who asserts ~α is truly committed to the rejection of α: asserting α as well would commit one to triviality. (p. 222. The logical symbols are mine.)

On the other hand, a paraconsistent logic, such as, Batens argues, does not allow us to "express correctly that we reject some sentence" (pp. 222-3). For asserting both α and ¬α does not lead to triviality.[23]

---

[21]Of course, ~ and ¬ are 'compatible' in the sense that they can both occur in the same logic, as is shown by Meyer and Routley (1973) and Meyer and Routley (1974). But the discussion here is concerned with the question of whether or not ¬ is too weak for some context.

[22]Then *non-primeness* (a theory Σ is non-prime iff for some sentences α and β, α ∨ β ∈ Σ but α 6 Σ and β 6 Σ) is also rejected. For primeness follows from consistency and completeness (given De Morgan's laws). See Mortensen (1983) pp. 37f.

[23]The same objection is applicable to a theory of belief revision whose underlying logic is paraconsistent. Batens (1980) argues that "only those theories are informative that «forbid» something". (p. 227) Yet in a theory whose underlying logic is paraconsistent, "no sentence will ever lead to the rejection" (p. 231). So if we use a belief set to represent one's beliefs and use a paraconsistent logic as an underlying logic, then there is nothing in the theory of belief revision that commits us to revise our beliefs. In reply to this problem, even if one rejects and so there is no logical reason to revise our beliefs, when the beliefs become *incoherent*, one may revise their beliefs. See Tanaka (1995) and Tanaka (1998) for this line of reply. And beliefs being

It is not clear whether the exclusion of α by its negation is an element of the realm of . The exclusion may be presupposed as a norm that governs reasoning in some context. If a context is a communication situation, we may agree to exclude something by asserting its negation. In other words, the exclusion can be a feature of a logical system as a theory that we use in some context. Yet it is questionable whether the exclusion is part of the 'structure of norms' which seems to be what the Australian school takes to be, as can be seen from Priest quoted above. Just as it is absurd to argue that the structure of a building is itself the building, it does not seem reasonable to suppose that the structure of norms *is itself* a norm. Even if it is argued that the exclusion is part of ,[24] there is no reason why paraconsistent logicians cannot introduce into the language of the logical system an absurdity constant, $\perp$ (or f), and let it be governed by a rule $\perp \vdash \beta$ for all β. In order to express the exclusion of α by ¬α, one may further introduce the rule $\{α, ¬α\} \vdash \perp$. In this way, asserting both α and ¬α leads to triviality. In effect, thus, we have . Therefore, that α rules out ¬α can be expressed paraconsistently as in classical logic, despite the claim made by Batens.

It may be argued that what encapsulates is the functionality of evalua-tions of formulas. If an evaluation is a function, every formula is assigned either true or false, but never both (and never neither). So, it may be argued, what expresses the exclusion of α and its negation is the definition of ~ together with the functionality of evaluations. And paraconsistent logics, in particular, whose evaluation of a formula is a relation instead of a function, do not capture the spirit of the exclusion, although they can introduce some rules in order to 'imitate' its effect.

None the less, the exclusion can be expressed in a paraconsistent logic with the introduction of the two rules mentioned above. If the exclusion is an element of and so has to be captured by a logical system as a theory, the rules serve to express the exclusion in the system. Of course, we need to debate what is the best way to express it. The functionality of evaluations of formulas together with the definition of ~ may well give rise to a 'better' theory. In any case, the fact that the exclusion can be expressed paraconsistently undermines Batens' claim that paraconsistent logics are too weak to be applicable in some context.

To sum up: we have considered the Batens objection that there is some context in which paraconsistency fails, in particular, ¬ is weaker than ~. How-ever, it does not seem that Batens has made his case.

---

incoherent may have nothing to do with logic. There may be many *a posteriori* reasons why beliefs are incoherent. See Priest (2001) for an account of belief revision which is based on this idea.

[24]Note that if the exclusion is not an element of the realm of , classical logic, that has that element, does not capture properly.

7. SUMMARY

I have argued that the objections raised by the Belgian and Brazilian schools of paraconsistency to the Australian school are not well founded. This is not to argue that the Australian approach is justified. There may be legitimate objections that could not be rejected. None the less, the Brazilian school's objection is based on a mis-conception of the Australian school. And the Belgian school's objection is not successful. Thus the Australian school need not be closed down on the basis of Belgian and Brazilian schools' objections.

8. APPENDIX

Having dismissed the Belgian and Brazilian schools' objections to the Australian school, let's now carefully examine the Brazilian school's view of logic.[25] It does not seem that their view is coherent.

8.1 AGNOSTICISM, PLURALISM, AND THE BRAZILIAN SCHOOL OF PARACONSISTENCY

In advancing agnosticism in the study of logic, da Costa and Bueno (1996) reject the notion of 'true logics'. They write:

If they [viz., applied logics] are to be minimally successful, perhaps our «radical» might claim, they have to be true, at least as far as their domains are concerned. This is an interesting remark. The problem underlying it, as in general with any radical view, consists in supplying evidence to the claim that such logics are in fact true. No means though seem to be available to offer such an evidence (there seems to be a considerable underdetermination at this level). (p. 55)

Both the classical school and the Australian school of paraconsistent logic have independently given some evidence as to why their logic is the true logic, al-though they argue for different logics on the basis of different evidence. Unless da Costa and Bueno can successfully reject the evidence given by these schools, their argument can hardly be taken seriously.

Moreover, da Costa and Bueno argue that a logic has its particular domain: an all-embracing logic, appropriate to all domains is hard to find. We are thus in general left with (several) alternative logics that describe only some aspects of them, and there are many heuristic and pragmatic reasons to choose between such logics, de-

---

[25]The following consideration is not applicable to the Belgian school. In this appendix, I consider only the Brazilian school.

pending in particular, of course, on the specific traits found in such domains. (p. 54)[26] However, if it is a problem to supply evidence to justify a logic being true, it must also be a problem to provide heuristic and pragmatic reasons for the claim that a logic has its own domain. For, unlike the instrumentalist, da Costa and Bueno do not ignore the realm of completely. Da Costa and Bueno are required to provide some reasons why it is 'this' logic instead of 'that' logic that has this particular domain. Hence they have to be concerned with the true logic for each domain. However, that is the concern that they identify as a problem in the study of logic.

None the less, it is not clear whether the Brazilian school subscribes to realism or instrumentalism.[27] On the one hand, they reject realism by not being concerned with in their study and development of logic. On the other hand, they reject instrumentalism by considering 'domains' which are the realm of . It may be possible to establish a middle ground between the two.[28] Yet their view of logic has not been well enough articulated to do so. It seems that the Brazilian school's view of logic is not very coherent.

9. REFERENCES

Arruda, A.I. (1989) 'Aspects of the Historical Development of Paraconsistent Logic', *Paraconsistent Logic: Essays on the Inconsistent*, G. Priest, R. Routley, and J. Norman (eds.), Philosophia Verlag, München, pp. 99–130.

Batens, D. (1980) 'Paraconsistent Extensional Propositional Logics', *Logique et Analyse*, Vol. 23, pp. 195–234.

Batens, D. (1985) 'Meaning, Acceptance, and Dialectics', *Change and Progress in Modern Science*, J.C. Pitt (ed.), Dordrecht, D. Reidel Publishing Company, pp. 333–360.

Batens, D. (1989) 'Dynamic Dialectical Logics', *Paraconsistent Logic: Essays on the Inconsistent*, G.Priest, R.Routley and J.Norman (eds.), Philosophia Verlag, München, pp. 187–217.

Batens, D. (1990) 'Against Global Paraconsistency', *Studies in Soviet Thought* Vol. 39, pp. 209–229.

---

[26] They continue: "Thus the relativist threat, based on the claim that there are no criteria of choice between rival logics, can be at least in part circumvented" (p. 54). However, the relativist does not claim that there are no criteria of choice. It is just that the criteria are relative to some factors. Thus it is not clear how da Costa and Bueno do not count as relativists.

[27] Priest (2000) points out that da Costa (1997) is confused about realism and instrumentalism.

[28] For example, Detlefsen (1986) takes Hilbert's programme to lie between realism and nomi-nalism (instrumentalism).

Batens, D. (1992) 'Do We Need a Hierarchical Model of Science', *Inference, Explanation, and other Frustrations: Essays in the Philosophy of Science*, John Earman (ed.), Berkeley, University of California Press, pp. 199–215.

Beall, J.C. and G. Restall (2000) 'Logical Pluralism', *Australasian Journal of Philosophy*, Vol. 78, pp. 475–493.

da Costa, N. C. A. (1997) *Logiques classiques et non classiques*, Masson, Paris.

da Costa, N. and O. Bueno (1996) 'Consistency, Paraconsistency and Truth (Logic, the Whole Logic and Nothing but 'the' Logic)', *Ideas y Valores*, No. 100, pp. 48–60.

Detlefsen, M. (1986) *Hilbert's Program: an Essay on Mathematical Instrumentalism*, D. Reidel Publishing Company, Dordrecht.

Haack, S. (1978) *Philosophy of Logics*, Cambridge University Press, Cambridge.

Haack, S. (1996) *Deviant Logic, Fuzzy Logic*, The University of Chicago Press, Chicago.

Jaskowski,´ S. (1948) 'Rachunek zdah dla systemów dedukcyjnych sprzecznych', *Studia Societatis Scientiarun Torunesis*, Sectio A, Vol. I, No. 5, pp. 55–77, reappeared as 'Propositional Calculus for Contradictory Deductive Systems', *Studia Logica*, Vol. XXIV, pp. 143–157.

Jenning, R.E. and P.K. Schotch (1981) 'The Preservation of Coherence', *Studia Logica* Vol. XLIII, 1/2, pp. 98–106.

Meyer, R. and R. Routley (1973) 'Classical Relevant Logics', *Studia logica*, Vol. 31, pp. 51–68.

Meyer, R. and R. Routley (1974) 'Classical Relevant Logics II', *Studia logica*, Vol. 33, pp. 183–194.

Mortensen, C. (1983) 'The Validity of Disjunctive Syllogism Is Not So Easily Proved', *Notre Dame Journal of Formal Logic*, Vol. 24, pp. 35–40.

Mortensen, C. (1995) *Inconsistent Mathematics*, Kluwer Academic Publishers, Dordrecht.

Parsons, T. (1990) 'True Contradictions', *Canadian Journal of Philosophy*. Vol. 20, pp. 335–353.

Perzanowski, J. (1997) 'Parainconsistency or Inconsistency Tamed and Investigated', a paper read to the First Congress on Paraconsistency, Gent.

Priest, G. (1979) 'The Logic of Paradox', *Journal of Philosophical Logic*, Vol. 8, pp. 219–241.

Priest, G. (1987) *In Contradiction: a Study of the Transconsistent*, Martinus Nijhoff Publishers, Dordrecht.

Priest, G. (1989) 'Reductio ad Absurdum et Modus Tollendo Ponens', *Paraconsistent Logic, Essays on the Inconsistent*, G. Priest, R. Routley and J. Norman (eds.) Philosophia Verlag, München, pp. 613–626.

Priest, G. (1999a) 'Validity', *The Nature of Logic* (European Review of Philosophy, Volume 4), A. Varzi (ed), CSLI, Stanford.

Priest, G. (1999b) 'What Not? A Defence of Dialethic Theory of Negation,', *What is Negation?*, D.M. Gabbay and H. Wansing (eds.), Kluwer Academic Publishers, Dordrecht, pp. 101-120.

Priest, G. (2000) 'Review of 'N.C.A. da Costa *Logiques classiques et non classiques*', *Studia Logica*, Vol. 64, pp. 435-443.

Priest, G. (2001) 'Paraconsistent Belief Revision', *Theoria*, Vol. 67, pp. 214-228.

Priest, G. (2002) 'Paraconsistent Logic', *Handbook of Philosophical Logic* (Volume 6, Second Edition), D. Gabbay and F. Guenthner (eds.) Kluwer Academic Publishers, Dordrecht, pp. 287–393.

Priest, G. and R. Routley (1989a) 'Systems of Paraconsistent Logic', *Paraconsistent Logic: Essays on the Inconsistent*, G. Priest, R. Routley and J. Norman (eds.), Philosophia Verlag, München, pp. 151–186.

Priest, G. and R. Routley (1989b) 'First Historical Introduction: a Preliminary History of Paraconsistent and Dialethic Approaches', *Paraconsistent Logic: Essays on the Inconsistent*, G.Priest, R.Routley and J.Norman (eds.), Philosophia Verlag, München, pp. 3-75.

Priest, G. and K. Tanaka (1996) 'Logic, Paraconsistent', *Stanford Encyclopedia of Philosophy* (http://plato.stanford.edu/), Centre for the Study of Language and Information, Stanford University, Stanford.

Quine, W.V.O. (1970) *Philosophy of Logic*, Prentice-Hall, Englewood Cliffs.

Rescher, N. and R. Brandom (1979) *The Logic of Inconsistency*, Rowman and Littlefield, NJ.

Rescher, N. and R. Manor (1970) 'On Inference from Inconsistent premisses', *Theory and Decision*, Vol. 1, pp. 179–217.

Restall, G. (1994) *On Logics without Contraction*, Ph.D Thesis, University of Queensland.

Restall, G. (2000) *An Introduction to Substructural Logics*, Routledge, London.

Routley, R. (1980) *Exploring Meinong's Jungle and Beyond*, Research School of Social Sciences, Australian National University, Canberra.

Schotch, P.K. and R.E. Jennings (1980) 'Inference and Necessity', *Journal of Philosophical Logic*, Vol. 9, pp. 327–340.

Sylvan, R. (1989) *Bystanders' Guide to Sociative Logics*, Research School of Social Sciences, Australian National University, Canberra.

Sylvan, R. (1997) *Transcendental Metaphysics*, The White Horse Press, Cambridge.

Tanaka, K. (1995) *Paraconsistent Belief Revision*, Honours Thesis, University of Queensland.

Tanaka, K. (1998) 'What Does Paraconsistency Do? The Case of Belief Revision', *The Logica Yearbook 1997*, T. Childers (ed.), Filosophia, Praha, pp. 188–197.

Funny enough is the fact that I, during what I call my martyrdom, which are the years that go from end of 2001 until today, and after today, ended up doing my best to defend Priest's thinking from back then, so that the piece you will see now is my best attempt of making Ontological Paraconsistency become something plausible in Science.

By the time I did this, this was my best attempt of pleasing Priest. I am doing this to try to have him helping me get a permanent academic position somewhere, since I am feeling as if I am the most violated of the creatures, the most attacked, and I am seeing no way out. I realistically put all my mental powers to work in that direction, so into changing what I initially thought was laughable into something scientifically acceptable. See what I came up with, please:

International Journal of Philosophy
2016; X(X): XX-XX
http://www.sciencepublishinggroup.com/j/ijp
doi: 10.11648/j.XXXX.2016XXXX.XX
ISSN: 2330-7439 (Print); ISSN: 2330-7455 (Online)

Ontological Paraconsistency Has a Place

Marcia Ricci Pinheiro

IICSE University, Wilmington, USA

Email address:
drmarciapinheiro@gmail.com

To cite this article:
Marcia Ricci Pinheiro. Ontological Paraconsistency Has a Place. *International Journal of Philosophy*. Vol. x, No. x, 2016, pp. x-x.
doi: 10.11648/j.xxx.xxxxxxxx.xx

Received: MM DD, 2016; **Accepted**: MM DD, 2016; **Published**: MM DD, 2016

Abstract In this paper, we recover the idea cast by Graham Priest to our ears in 2000: That it was possible to experience Ontological Paraconsistency in life. He had, back then, as a translation of his thinking, a painting by Escher: The stairs could be going up or down, and one could not tell where they were going by simply examining the painting. The most obvious argument as to why that was not an instance of Ontological Paraconsistency found in reality was that the perspective from which you would have to stare at the painting to see something different would be different too, so that it was impossible that we were getting up and down at the same time, that is, from the same perspective. That would happen with anything we picked in this world. We recently found something that does not entirely belong to this world, however, something that could finally satisfy the requirements of Priest, and therefore prove to us that there is a place for Ontological Paraconsistency. We observe that the paraconsistent robot, Emmy (Abe et al., 2006), is an application of the Non-ontological Paraconsistency, which we always believed to be passive of existence, but we here talk about another type of paraconsistency, which would be intrinsic to the being. The purpose of this paper is then providing a definite answer to the questions: Is there any real life instance of Ontological Paraconsistency? Is Ontological Paraconsistency a useful concept in terms of logical theories?

Keywords: Paraconsistency, Priest, Da Costa, Ontological, Tanaka, Non-ontological, Classical Logic, Escher

International Journal of Philosophy

2016; X(X): XX-XX

http://www.sciencepublishinggroup.com/j/ijp

doi: 10.11648/j.XXXX.2016XXXX.XX

ISSN: 2330-7439 (Print); ISSN: 2330-7455 (Online)

1. Introduction

The difference between intrinsic and extrinsic paraconsistency, or between ontological and non-ontological paraconsistency, is that one brings, for instance, inconsistencies to our perceptions as a set of paradigms, so that we may believe that something is round and also square, say when people were yet to start spatial explorations and some theories said that the world was a cube or something similar to that whilst others said that it was a ball or something similar to that (there is a conflicting pair of members in our beliefs set, if human kind is considered as one thing, since we believe that something is rounded but we also believe that it is not rounded, percentages and all else ignored) and the other brings inconsistencies to the own being, which sometimes may be us.

When we say that the inconsistency is intrinsic, we could have the same individual using the same paradigms, and reaching different conclusions or the object being round and square at the same time, not our perception of it.

Our point, which we made clear to Priest in 2000, is that, when the individual says, *this is X*, they have a set of paradigms W in their mind, but, when they say, *this is not X*, they have another set of paradigms, so say W'.

With this, the object of observation is not both X and not-X, it is something independent, and we talk about perceptions and different paradigms each time.

With Emmy or any other robot made out of Paraconsistent Logic, there would be inconsistencies in the perceptions of the robot, and those can now be addressed. They were not being addressed before, however, not with Classical Logic. Now, it suffices programming the robot to perform action Y instead of crashing when conflict arises. All we are doing is avoiding the Law of the Explosion in this case.

We see the object as something invariably connected to our perception, that is, the object does not exist per se as something. We give a name to the perception paradigms of the vast majority when observing that object, so say moon, but that does not mean that everyone on earth has agreed that that object is the moon. Some may think it is cheese, for instance, and never drop that belief.

A person who loves another, so say M loves N, may also not love, that is, M may not love N, and that is common in human kind. For instance, the same guy who married a woman at the Catholic Church and promised, before the priest and the church attendees, to love that woman forever and ever may kill her all of a sudden. The thing, however, is that this guy does not love and love at the same time or because of the same reasons, and that is obvious: Perhaps he wants to kill her because he believes she betrayed him and he loved her because he believed that she was beautiful. She was beautiful and he loved her on a Monday. She was a traitor and he killed her on a Friday.

In this way, the conflicting feelings of this individual do not make him inconsistent and the own feelings are not inconsistent either if put together with the context, which is what we always have to do to analyse real life situations well.

The Sorites could be a problem for Classical Logic if out of context, if totally modified, as we explained in (Pinheiro, 2016), but it is not a Classical Logic problem inside of the context in which it is originally presented.

Real life makes a lot of difference and we explain that also in Words for Science (Pinheiro, 2015b).

How can a person be black and white? Say Michael Jackson is black and white like he himself says. He is black because he was born black, but he is white because he acquired the colour white. Once more, the moments in time are distinct and so are the paradigms involved: When we say that he is black, we think of his birth and the aspect of his skin. When we say he is white, we think of the moment of his death and the aspect of his skin.

Even though *aspect of skin* is seen on both occasions, the other paradigm is different, providing us with W and W' instead of W and W, what then explains why this is not an example of application of Ontological Paraconsistency, only at most an application of Non-ontological Paraconsistency.

In thinking like that, we concluded that nothing that we see as normal could fit the concept of Ontological Paraconsistency, but something like God could: Something mystical or almost purely abstract.

2. About God and Paraconsistency

With God, our beliefs are inconsistent at least sometimes if considered in isolation: He is good because He helped me win here. He is bad because He made me lose there.

The thing is that the own God is inconsistent if we consider that God is, for instance, His told-to-be declarations, what we see in the Catholic Bible: He says, *Do not kill* (Mark, 2016). He also says (in other words), *Sacrifice sheep for me* (Exodus, 2016).

That is clearly Kill and Do Not Kill because we don't know what He was thinking when He said Do Not Kill, and therefore it may as well be that it was about sheep and now we have both Kill and Do Not Kill with precisely the same contextual supplement, all coming from God.

Some think that God is a person and the Bible seems to deal with God like that: We are the image of God.

Some think that God is energy and materializes Himself as whatever He likes, so that He appeared in the skies for Jesus, but appeared in person in the shape of Jesus to others.

The beliefs of people are inconsistent in what regards God, but, once more, the paradigms are different: We have W and W' when that happens.

With the same people observing, however, and with those people having the same beliefs, so say no pre-established belief, and the same God, we could have God in the skies and God as Jesus, what then could make of God an example of application of Ontological Paraconsistency: He is Jesus but He isn't. When we see Him in the skies speaking to Jesus, then He isn't. When we see Jesus and believe that He is God, then He is.

Our set of beliefs could be that God is the unknown. In this case, God is both Jesus and the thing in the skies speaking to Jesus.

When we see a certain book and say, *It is English, but it isn't*, we may imply that the language of the book is definitely English, but it is impossible to understand it because it is Ancient English. We then mean Modern English when we said English and Ancient English is not Modern English, so that this is another language in that sense. It is still all about English and we do have the same paradigm in our mind when we say that, right? Notwithstanding, when we say Ancient English, we have the paradigm *ancient words* in our heads and when we say English, we have the paradigm *modern words* in our heads, so that this is still about different things, not Ontological Paraconsistency, since Ontological Paraconsistency happens neither inside of us nor inside of the book.

That must be why they created the concept of Holy Trinity: Father, Son, and Holy Spirit.

It seems that, at the same time, with the same paradigms, we see God as all of them.

3. About Escher and Paraconsistency

Fig. 1. Esher's Relativity (Shivprasad, 2011).

The picture that you have just seen portrays a famous painting of Esher that Graham Priest used to explain Ontological Paraconsistency better in that 2000 at the

University of Queensland (RIES, 2013).

In the painting, things would both be and not be, which should then make us believe that Ontological Paraconsistency is a reality.

Shivprasad (2011) talks about Relativity in the following way:

In *Relativity* Escher plays with our orientation of dimensions. We just cannot be sure where the ground is and where the sky is. The feet of the characters in *Relativity* could well be planted in the sky but one man's sky is the other man's ground: another paradoxical thought. *Relativity* is portrayed extensively in a succession of scenes in which Cobb introduces Ariadne to the basics of creating architecture for a dream. When they are walking the streets of Paris, the ground folds up, and becomes the sky. Buildings are seen inverted. Roads with cars running on them are seen in the sky. When Cobb and Ariadne begin walking, they climb up vertical roads and walk just as they would on the ground, just like the characters from *Relativity*.

Esher himself says (Kammer, 2014), in 1963:

I cannot help mocking all our unwavering certainties. It is, for example, great fun deliberately to confuse two and three dimensions, the plane and space, or to poke fun at gravity. Are you sure that a floor cannot also be a ceiling? Are you absolutely certain that you go up when you walk up a staircase?

Fig. 2. Observe the guy that carries the candle. He seems to be going upstairs, right? The two guys to the right: One seems to be going downstairs and the other seems to be starting to hold the support, so that he can go upstairs or even downstairs after he succeeds.

Fig. 3. *The guy who carries the candle looks as if he is going downstairs. The two guys to the left look as if they are going upstairs in a weird manner.*

The basic problem, as explained before, is that, when we see the guy with the candle going upstairs, the angle of our sight is W, let's say, but, when we see him going downstairs, the angle is W', what we noticed in the own 2000. In this case, it is not the same things that both are and are not, since our angle of sight has changed, and therefore our mental paradigms. Things present themselves in a way X from angle W and in a way X' from angle W'. Then one could claim that they are paraconsistent, and Priest would probably say that they are.

We would say that a person of 1.72m may look tall if put together with a pigmy and short if put together with a giant, say a guy who is 2.5m tall. It is the same person and that does not make them short and not-short at the same time: They are relatively short at a time Y and relatively tall at a time U, different from Y, and therefore not at the same time. The perspective of those observing has also changed: First, they had the extremely short person to compare the basic height with, then they had the extremely tall person. The height of the person being observed, however, never changed: What changed was the perspective of those observing due to paradigms that were introduced around the person. In a certain town, we could say that a man of 1.72m has a regular height, so that his height is always regular, never high or low, regardless.

As we turn the picture of the painting, the angle of observation changes as well as what we see close to any selected point, which is the same that happens to the person whose height was observed in the previous paragraph, and therefore those selected points have not changed and are not paraconsistent entities.

One could also argue that this is the same we see in the example with God and Jesus: When we observe Him with Jesus, we see things from a certain perspective and together with a set of paradigms, say W. When we observe Jesus on his own, we see things from another perspective together with a new set of paradigms, say W'. Notwithstanding, our idea of what God is in our case, *the unknown*, has not changed, and therefore our basic paradigm has never changed. We see Jesus and his shape with no mistake and no change in what we see, regardless of the perspective, different from what happens with Escher's painting. We see God and His shape with no mistake and no change in what we see, also regardless of the perspective. What is happening here is that both will fit our definition of God with no mistake and in the same way. In this case, we are obliged to agree that God both is and isn't Jesus. Even so, before we see God speaking to Jesus from the clouds, God is Jesus and Jesus only. It is only after we see God speaking to Jesus in that way that we think God may be something else or **also** something else, and perhaps Jesus is a human like us compared to that something else. We have more context, more information, and therefore other paradigms ALSO here. Jesus is then God for us, but perhaps he is a lesser god or something not as unknown as the other god. These paradigms, the ones that differ here, would be something like revealing more pieces of the painting of Escher, so say they found out that part of his painting was actually hidden behind the frame and now we have more data. This is something completely different from simply turning the picture and seeing from another angle. Yet, this ALSO promotes a change of paradigms that is non-negligible.

The thing is that if it all depends on our observation, what is inside of our minds, then it is not the object of our observation that is inconsistent, but at most our private logic. As our private logic could be described by a special logical system, this paraconsistency is non-ontological.

We can draw a set and call it Unknown. In this set, we may insert Jesus. Now the idea is clear: Unknown is Jesus, since Jesus is inside of Unknown. Notwithstanding, Unknown is not Jesus because Unknown is more than Jesus. Once more, we have the perspectives being different, right? In one case, we only see Jesus inside and we then say that Unknown is Jesus. In the other case, we see the rest, perhaps finally, then Unknown is not Jesus or Unknown is not only Jesus.

If we consider that different people have different beliefs about God, however, and we take them all to be true because, for instance, they all seem to have passed through miraculous processes, God could end up being a person and a non-person, so say we take the Catholic belief, that Jesus was God, and therefore human during His life on earth, and the belief of The Empire (Pinheiro, 2015), that Jesus was just a normal human and, if God has something to do with him, then He has to do with us in the same way. The Empire believes that God was never human.

In this case, God is both a human and a non-human and He is then paraconsistent.

Each religion that we have just mentioned has their own perspective, but our perspective, of those observing them, is the same: They have miracles to present and therefore their gods should all be our God. In this case, God is both human and non-human at the same time. Even though our explanation for that is that those religions say that He is such and such and they have such and such miracles, God must still satisfy all that to be the only god we have, and this was our decision here. In this case, He must be both human and non-human, therefore paraconsistent in what regards at least this aspect.

Perhaps it is worth examining a few more pictures:

Fig. 4. *Duck Rabbit Illusion (Illusionist.com, 2015).*

In this picture, we should see a duck if we look at it from right to left and a rabbit if we look at it from left to right. It is probably clearer here: It all depends on the angle of our sight, and therefore the conflicting impressions do not come from the same origin, what then makes each one of the impressions happen at a different time, not at the same time, as it would be essential for us to claim that this is a real-life instance of the phenomenon Ontological Paraconsistency.

Fig. 5. *Candlestick Illusion (Illusionist.com, 2015).*

In this picture, we should see two people if we fixate our eyes on the white part and a candlestick if we fixate them on the black part. Once more, things do not happen at the same time. Maybe, however, for other types of entities, it would be possible to see both at the same time. Maybe there are special illnesses or genetic modifications that would allow our eyes to see both images, faces and candlestick, at the same time. In those special cases, it would be possible to assert that this is an instance of Ontological Paraconsistency, finally, like if our eyes never moved and our thinking never changed, then we could probably say that it is if we see both at the same time. In human race as we know it, however, this is not an instance of Ontological Paraconsistency.

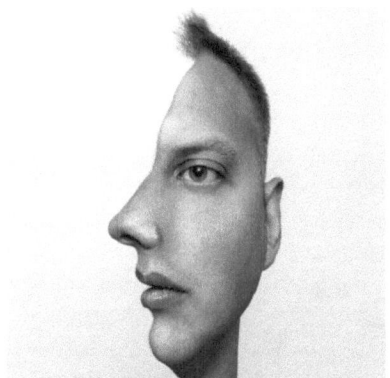

Fig 6. *Double Face (Pixdaus Ltd., 2016).*

Here, what we see is a frontal picture of half of a face if we fixate the eyes on the ear of the character and a full picture, of the type caricature, of the side of a face if we fixate our eyes on the nose and mouth that we see to the left. As just said, the eyes go different places to see different things, so that this is NOT an example of Ontological Paraconsistency.

4. Conclusion

We seemed to be stuck with concluding that there is no such a thing as Ontological Paraconsistency, that all that there is is Non-ontological Paraconsistency, as Da Costa apparently believes, this according to what Priest said by means of personal communications in 2000 to us, and according to what Tanaka said in (Tanaka, 2003). Tanaka replaces the term Ontological Paraconsistency with perhaps an expression, *accepting of true contradictions. True contradictions* is a term that obviously implies contradictions that exist as a fact in the world. If they exist as a fact, then things are ontologically paraconsistent, which is our choice of terms here.

It seems that whatever thing we attempt to put in the bag of Ontological Paraconsistency will end up either presenting different paradigms, as in W and W', in the analysis when the conclusion is that it should be or will end up allowing for the creation of a logical system, what then makes it all be non-ontological.

Escher's painting is a true inspiration but it more represents a puzzle than an example of Ontological Paraconsistency: It is a paradox up to the point at which we find out that there has to be a change in perspective and even in the context for us to end up asserting two conflicting things about the items in the painting.

Not even God suffers from Ontological Paraconsistency because the perspectives and contexts will be different when we reach conflicting conclusions. There is a case, however, a case involving us defining what God is based on what is stated by human kind, that proves that God may suffer from Ontological Paraconsistency: In this case, we have a group of humans and we say that we must have only one god. All their definitions of God, or all their beliefs about God, would have to be true at the same time. That is when we could say that God both is and isn't something, so say human.

The difference between this example and the situation in which we have The Bible in front of us and a group of people saying that it is blue whilst another group is saying that it is green is that we ourselves are going to say that it is either blue or green when somebody asks us what we think about it. Now, if we say that it is both green and blue, then we are actually seeing green from one perspective, a set of paradigms W, and blue from another perspective, a set of paradigms W', what then does not give an example of Ontological Paraconsistency, just confusion in the description of our observations/perceptions. We can obviously write a code to explain that, that is, we can express that by means of a programming language. Whatever fits a programming language is part of the phenomenon Non-ontological Paraconsistency instead. It is not the object that is both blue and green, but our perception that is confusing, our association between sigmatoids and world objects, that is, it is our perception that is not being well described. Notwithstanding, it could be perfectly described by means of the human language too.

Perhaps any non-animated object would have to not suffer from Ontological Paraconsistency, since it actually depends on us acknowledging its existence to become something, and something we invent, with our abstract world, that they are. In this case, what we invent that they are depends on our perception, our human perception, and therefore the only entities that could suffer from Ontological Paraconsistency would be us. In this case, we would have to be and not to be at the same time, considering the same perspective or the same paradigms, for us to be able to say that we are Ontologically Paraconsistent. The example we here mention, about the human marriage, should make this all clearer.

References

[1] Exodus 20_24 KJV - An altar of earth thou shalt make unto - Bible Gateway. (n.d.). Retrieved February 29, 2016, from https://www.biblegateway.com/passage/?search=Exodus+20:24&version=KJV

[2] Illusionist.com. (2015). Double Meanings. Retrieved March 26, 2016, from http://www.optical-illusionist.com/category/double-meanings/

[3] Inácio, J., Torres, C. R., Abe, M., & Filho, D. S. (2006). Robô Móvel Autônomo Emmy (Autonomous Mobile Robot, Emmy): Uma Aplicação eficiente da Lógica Paraconsistente Anotada (An efficient application of the annotated paraconsistent logic), 19–26.

[4] Kammer, C. (2014). Esher's impossible stairs inspired by high school stairwell. Retrieved March 26, 2016, from http://www.nrc.nl/nieuws/2014/11/16/eschers-impossible-stairs-inspired-by-high-school

[5] Mark 10_19 KJV - Thou knowest the commandments, Do not - Bible Gateway. (n.d.). Retrieved February 29, 2016, from https://www.biblegateway.com/passage/?search=Mark+10:19&version=KJV

[6] Pinheiro, M. R. (2015). (August 2015). The Empire of God.

[6] Retrieved February 28, 2016, from http://theempireofgod.blogspot.com.au/2015_08_01_archive.html

[7] Pinheiro, M. R. (2015b). Words for Science. Indian Journal of Applied Research, 5(5), 19–22. Retrieved from https://www.worldwidejournals.com/ijar/articles.php?val=NjQ0MQ==&b1=853&k=214

[8] Pinheiro, M. R. (2016). FIRST DECISIONS: NATURE OF THE MODELLING WORK.

[9] Pixdaus Ltd. (2016). Pixdaus. Retrieved March 26, 2016, from http://pixdaus.com/double-face-photo-by-jdtnt-illusion-appearing-as-two-faces/items/view/579957/

[10] Priest, G., Tanaka, K., & Weber, Z. (2013). Ecology (Stanford Encyclopedia of Philosophy). Retrieved from http://plato.stanford.edu/entries/logic-paraconsistent/

[11] RIES RAMON LLULL. (2013). escalator « Bla, bla, bla 3. Retrieved February 27, 2016, from https://1eso1314.wordpress.com/tag/escalator/

[12] Shivprasad. (2011). Inception and its Inspirations: Esher, Dali and Yoga Vasistha. Retrieved March 26, 2016, from http://www.criticaltwenties.in/philosophyreligionculture/inception-and-its-inspirations-escher-dali-and-yoga-vasistha

[13] Tanaka, K. (2003). Three Schools of Paraconsistency. Australasian Journal of Logic, 1(1), 28–42. Retrieved from http://philosophy.unimelb.edu.au/ajl/2003/

International Journal of Philosophy
2016; X(X): XX-XX
http://www.sciencepublishinggroup.com/j/ijp
doi: 10.11648/j.XXXX.2016XXXX.XX
ISSN: 2330-7439 (Print); ISSN: 2330-7455 (Online)

Upon being requested by the group to which the above scientific journal belonged to, Science Publishing, to produce a marketing token for the paper, something that could attract people to read it if they read their newsletter, I produced the following text, which was never published because they gave up on publishing:

Do we have two blue concrete walls and a person comfortably walking over a surface that imitates a rock and then pictures or the sea and natural rocks and the person doing what is physically impossible?

This is a picture by Philippe Ramette found on http://houhouhaha.fr/philippe-ramette on the 1st of April of 2016

Ontological Paraconsistency and God: An Actual Possibility

What is Surrealism about? There is the feeling that it provokes, and there is the theory behind the art that is produced, right? It is about conflicts, surprises, and shocks, is it not?
http://www.arthistoryarchive.com/arthistory/surrealism/Origins-of-Surrealism.html **brings a bit about the theory behind the artistic movement Surrealism.**

What is Paraconsistency about? There is the feeling that it provokes when you see it in practice, and the theory behind it. It is about oppositions, contradictions, conflicts, and impasses.
https://www.youtube.com/watch?v=VbiMI2sZ3HU **brings images of Emmy, the Paraconsistent Robot. This is an example of the Non-ontological Paraconsistency because this is a computer system, basically: The entity itself is not paraconsistent, what is paraconsistent is the logical system that has been inserted in it.**

The most important icons of Paraconsistency in modern times would definitely be Graham Priest and Da Costa. They seem to oppose each other when one says that Paraconsistency is ontological and the other says that it is Non-ontological, but, in reality, they are simply exploring possibilities.

The author of this article had personal contact, face-to-face, in the condition of postgraduate research student, with Graham Priest in the year of 2000, and had at most virtual contact with a few members of Da Costa's team many years later. Graham Priest lectured her on his theory regarding the topic in the same year, 2000, via a discipline called Special Topics, which involved face-to-face weekly appointments.

The Non-Classical Logical Systems are a trial of subverting Classical Logic, so that each one of them denies some part of the Classical Logic Reasoning, which originally belongs to Mathematics and is used all over the world to prove or justify its foundations.

Paraconsistency appears as one more Non-Classical System and is usually seen as non-ontological.

It has been created by The Cologne School, perhaps, what means 15th Century (http://plato.stanford.edu/entries/logic-paraconsistent/#Par).

Ontology means the branch of Metaphysics that deals with the nature of the being (http://www.thefreedictionary.com/ontology). Ontological Paraconsistency then means Paraconsistency that is found in the nature of the being.

In this case, we look for conflicts in the own nature of the entity, so say *God is Jesus* and *God speaks to Jesus from the clouds* lead us to infer that *God is Jesus and God is not Jesus*, therefore that God is a conflicting entity, and its nature is confusing and contradictory. God then could suffer from Ontological Paraconsistency.

This is what Dr. Pinheiro tries to say in this little article: Maybe there is nothing that can be observed by humans that is not directly connected to how they see things and, if conflicts are found, then they are usually based on different mental paradigms of those humans that observe things. Notwithstanding, some things, which are not actually things, so say God, may give us different observational conclusions, conflicting ones, and still be based on the same mental paradigms, when we then would have an example of Ontological Paraconsistency, provided we accept that all still depends on the eyes of the beholder, basically: God Himself may never actually think that there is a chance that He is inconsistent, but our idea about Him might end up implying that, to us, He suffers from Ontological Paraconsistency.

Some relevant quotes from the sources of this article:

> A construcao e o desempenho do Robo Emmy veio comprovar que Sistemas de Controle totalmente baseados nos fundamentos da LPA sao possiveis de serem implementados. O robo Emmy, alem da importancia tecnica, traz uma efetiva contribuicao as pesquisas de Aplicacoes que procuram novos caminhos para tratar contradicoes atraves da utilizacao de Logicas Nao-Classicas no campo da Inteligencia Artificial.
>
> Robo Movel Autonomo Emmy: Uma aplicacao eficiente da logica paraconsistente anotada, Revista Selecao Documental, 03/2006 – J. I. S. Filho, C. R. Torres e J. M. Abe

> A logical consequence relation, ⊨, is said to be *paraconsistent* if it is not explosive. Thus, if ⊨ is paraconsistent, then even if we are in certain circumstances where the available information is inconsistent, the inference relation does not explode into *triviality*. Thus, paraconsistent logic accommodates inconsistency in a sensible manner that treats inconsistent information as informative. The prefix 'para' in English has two meanings: 'quasi' (or 'similar to, modelled on') or 'beyond'. When the term 'paraconsistent' was coined by Miró Quesada at the Third Latin America Conference on Mathematical Logic in 1976, he seems to have had the first meaning in mind. Many paraconsistent logicians, however, have taken it to mean the second, which provided different reasons for the development of paraconsistent logic as we will see below.
>
> Paraconsistent Logic – Encyclopedia of Philosophy, 2013 - G. Priest, K. Tanaka and Z. Weber

> In particular, the paper concerns three 'schools' of paraconsistency: Australian, Belgian and Brazilian.1 Arguably the most radical school, the Australian school of paraconsistency, led by Priest and Sylvan (né Routley),

claims that there are some true contradictions and that *the* logic is paraconsistent. (Sylvan did however advocate logical pluralism in a posthumously published work, Sylvan (1997).) The Belgian school, led by Batens, and the Brazilian school, led by da Costa, argue against the Australian school. They question the existence of true contradictions. More importantly, they not only reject the idea that *the* logic is paraconsistent but also deny that there is a uniquely correct logic.

Three Schools of Paraconsistency – Australasian Journal of Logic, 2003 – K. Tanaka

Author:

Dr. Marcia R. Pinheiro, IICSE University, DE, United States

Their invitation to write the above call for the article came via electronic letter. See:

From: Science PG <sciencepg.payment@gmail.com>
To: Marcia Pinheiro <mrpprofessional@yahoo.com>
Sent: Friday, 1 April 2016, 12:29
Subject: Re: Acceptance Letter of Paper [2041009] from SciencePG
Dear Marcia Pinheiro,
Thanks. we have received the payment for paper 2041009. You don't need to worry about the system status.

Later we will send the final edited paper to you for your final checking and confirmation before publishing, please note.

Now SciencePG opens a new column "SciencePG Frontiers" aiming to release our authors' research achievements in the form of news articles. So, we are writing to invite you to write a news for your article (it is different from your paper to be published. That is, to write your article into a readable news article. **It is free of charge.** When your article is published, the article news you write will be shown in "SciencePG Frontiers". In this way, more and more scholars will know your research achievements.

It is a rare chance to expand the influence of your research achievements. Therefore, hope you can offer your support, finish the news article and send it to us in your earliest convenience (within one week).

You can take a look at below link and attachments for details.
http://www.sciencepublishinggroup.com/news/sciencepgfrontiers

Looking forward to your positive reply.

Best wishes,
Iris White
Editorial assistant
Science Publishing Group, USA
http://www.sciencepublishinggroup.com
Email: sciencepg.payment@gmail.com

To the record, Doctor Professor Graham Priest had about 100 papers it the Philosophers' Index when I consulted it, and that was 2001. The Philosophers' Index is the most important search tool for articles in Philosophy I had contact with so far.

CHAPTER 2
John Corcoran

I met Professor Doctor John Corcoran through the Internet, a place called Academia.edu. He appears on Research Gate too. This is his profile there:

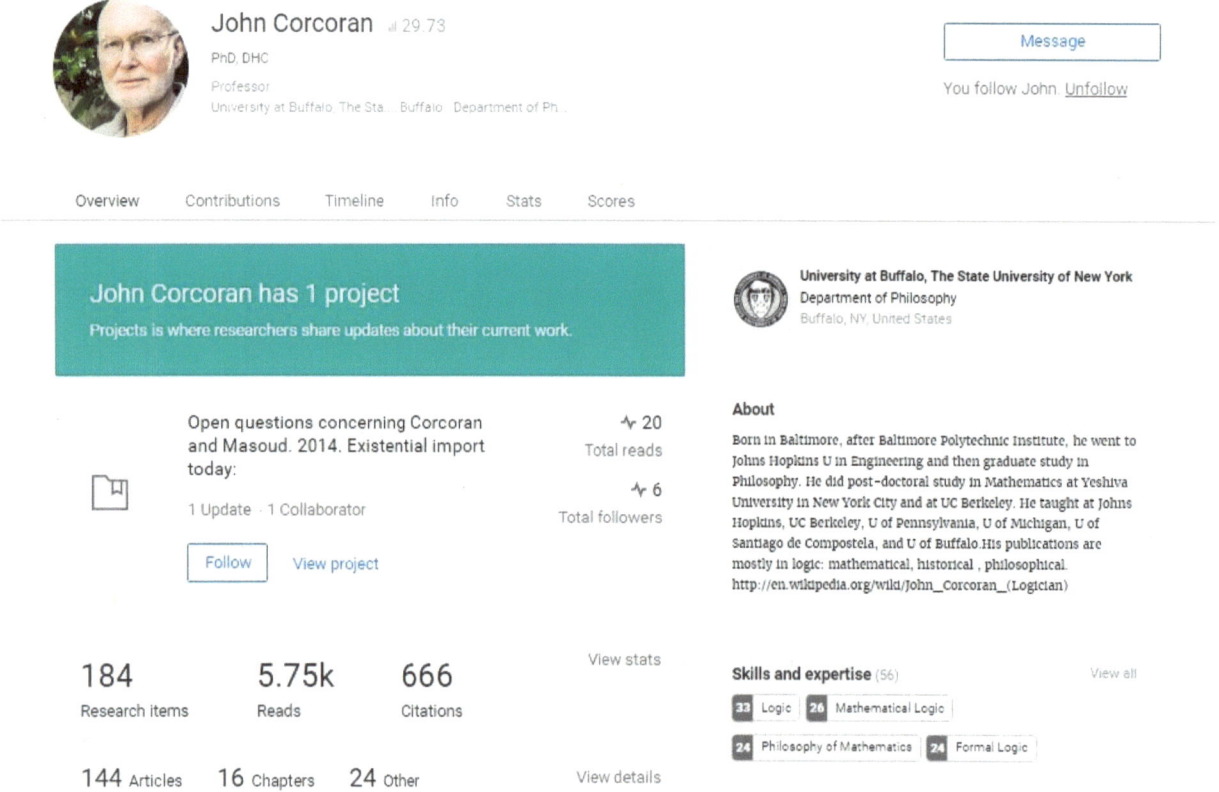

Peculiar is that 666, the number they classify as the number of the beast, appears in this picture.

That is the exact number of citations he has of his work in all that the tool, Research Gate, can see, let's say.

Professor Corcoran has been inspiring my writing for a while in recent times because he starts what is called Open Sessions at Academia.edu and he invites me to comb his work.

That is a very generous gesture in Science because every academic who is part of the mainstream has a choice: They can care about the opinion of others but they may also not care.

By offering me his paper, for me to read, and therefore criticize, Professor Corcoran is opening his heart and mind for my criticism of his work. That is a lot of a compliment, given that he is stating that he bothers about what I think in what regards his work.

He is not my supervisor or professor, so that he could never care about me, but he somehow found me and then asked: Marcia, what do you think?

These modern tools of Science, such as Research Gate and Academia.edu, are wonderful creations: We can get to know about the work of each other without knowing anything about the individuality of the other, that is, in a completely ethical way.

We can also end up establishing random partnerships, collaborations, and it is all because we actually read the work of the other.

Well, I was invited to read a few of his pieces, and, since more recently they finally let me keep my websites online (I was hacked in a continuous manner for more than 10 years, all ruined all the time in terms of whatever was IT), so that I ended up publishing a few tokens at WordPress and Blogger. Here you see the first piece about Corcoran's work.

The text you see below this line is a reproduction of a post I put online on 14/11/2016 with WordPress. See:

ON SELF-REFUTING: PROFESSOR CORCORAN

Recent philosophical literature discusses "self-refuting propositions" and "self-sustaining propositions". By one definition, a proposition "whose truth implies its falsity" is *self-refuting*; one "whose falsity implies its truth" is *self-sustaining*. Perhaps unsurprisingly, both proposition classes seem empty.

Let P be an arbitrary proposition. If P is true, then P's falsity does not exist. Thus, it is not the case that P's falsity is implied by its truth—something existent. Thus no true proposition is self-refuting. On the other hand, if P is false, then P's truth does not exist. Thus again, it is not the case that P's truth implies its falsity—something existent. Thus no false proposition is self-refuting. But this means that no proposition is self-refuting. Similar reasoning concludes that no proposition is self-sustaining.

(Corcoran, 2014)

The equivocated inference is this: If **P** is true, then **P**'s falsity does not exist. Thus, it is not the case that **P**'s falsity is implied by its truth-something existent.

If **P** is **This sentence is false**, some people would think that the veracity of **P** implies its falsity. In this case, **P**'s falsity may exist.

Perhaps Professor Corcoran confuses existence with veracity here. One concept differs fundamentally from the other however: If something does not exist, we cannot say it is true or false, for we cannot actually see it as existing, especially in language.

For us to evaluate **P** as false or true, **P** must exist, first of all. If **P** is false, **P** exists and is false. If **P** is true, **P** exists and is true. In both cases, **P** exists.

If **P** is true, **~P** still exists, since we can think of it and write about it. It is just that it is false, but this only in Classical Logic as well, since in other logical systems it is possible that **~P** be something else, say **0.5 true**.

As I keep on saying, these guys who thought of Classical Logic were realistically hardworking people, who really deserve our respect. The words they choose are as important as the concepts they create.

A reading that is worthwhile at this stage, for those who are interested in this topic, is **Words for Science**.

References

Pinheiro, M. R. (2016). On Self-Refuting: Prof. Corcoran. Retrieved 19 December 2016 from https://drmarciapinheiro.wordpress.com/2016/11/14/on-self-refuting-dr-corcoran/

2014. Self-refuting propositions and self-sustaining propositions. Bulletin of Symbolic Logic. 20 (2014) 250. (Coauthors: John Corcoran and José Miguel Sagüillo)

Pinheiro, M. R. (2015). Words for Science. *Indian Journal of Applied Research*, *3*(5), 19–22. Retrieved 19 December 2016 from https://www.worldwidejournals.com/ijar/articles.php?val=NjQ0MQ==&b1=853&k=214

We will usually find something useful in the work produced by logicians, so that it might be worth it reading all they write even when we stumble upon a piece that we think is absurd.

Professor Corcoran says: When the truth of a proposition implies its falsity, the proposition is self-refuting. If the falsity of a proposition implies its truth, however, it is self-sustaining.

Then he says that both classes of propositions seem to be empty.

I like the names self-refuting and self-sustaining.

I suggested, in the post: Make **P = This sentence is false**.

That is a classical example, taught to me by someone else a long time before I read Corcoran's piece.

If **P** is false, it must be true that the sentence is true. In this case, **P** would be **self-sustaining** according to the theory he presents to us.

If **P** is true, then **P** or this sentence is false. If we think that **P = this sentence**, then **P** is false. In this case, **P** is **self-refuting**.

Since **P** is both self-sustaining and self-refuting, it looks like these classifications are also not OK, unfortunately.

Since we were able to present at least one sentence that would be self-refuting and self-sustaining, these classes are not really empty.

The post you read below, which is about another topic, but is still about Professor Corcoran's ideas, was published on the 15//11/2016 on WordPress.

EXISTENTIAL IMPORT WITH TAUTOLOGY

My 2007 *"Existential import"* [1], which did not consider weak existential-import, proves the Strong Existential-import Equivalence [SEIE]: A universalized conditional has strong existential-import iff the existentialization of the conditional's antecedent predicate is tautological. Other proofs are in [2] and [3].

SEIE: $\forall x\ [S(x) \rightarrow P(x)]$ implies $\exists x\ [S(x)\ \&\ P(x)]$ iff $\exists x\ S(x)$ is tautological.

Thus as a corollary we have the Weak Existential-import Equivalence [WEIE]: A universalized conditional has weak existential-import iff the existentialization of the conditional's antecedent predicate is tautological.

WEIE: $\forall x\ [S(x) \rightarrow P(x)]$ implies $\exists x\ S(x)$ iff $\exists x\ S(x)$ is tautological.

However, many logicians will find it more convenient and perhaps more revealing to reverse the process: to prove the Weak Existential-import Equivalence first as a lemma and then get the Strong Existential-import Equivalence as a corollary.

[1] JOHN CORCORAN. *Existential import.* this BULLETIN, vol. 13 (2007). pp. 143–4.

[2] JOHN CORCORAN AND HASSAN MASOUD. *Existential-import today.* **History and Philosophy of Logic**. vol. 36 (2014). pp. 39–61.

[3] JOHN CORCORAN AND HASSAN MASOUD. *Existential-import mathematics.* this BULLETIN. vol. 21 (2015). pp. 1–14.

The issue here is the word tautological, for what is a tautology?

See the definition of Tautology:

> A **tautology** in logic is a formula that is always true on any valuation or interpretation of its terms. They are also sometimes called *valid formulas* (not to be confused with a valid argument) or *logical truths*

The most obvious and commonly used example of a tautology is the formula **A ∨ ¬A**. Under any valuation, whether **A** is true or **A** is false, **A** or **not-A** will always be a true statement.

We can easily verify that **A ∨ ¬A** is a tautology by means of a truth table:

A tautology may otherwise be defined as a formula that is satisfied under every possible valuation.

A	A ∨ ¬A
⊤	⊤
⊥	⊤

(Philosophy Index, 2016)

There exists **x** such as **S(x)** cannot be a tautology because if we evaluate that as false, then it is false, like it can be either false or true and there is nothing saying that the result of the totality of the assertion will be anything different from our choice for the parts, it is the opposite: There is only one declaration for us to assign truth-values to.

Professor Corcoran must then have referred to the entire assertion before that.

In this case, we have:

For all **x**, if **S(x)** then **P(x)** implies **there is x** such that **S(x)** and **P(x)** iff **there is x** such that **S(x)**.

Analysis:

An iff clause means if then from antecedent to consequent and vice-versa. Assuming the antecedent, which is, in this case, For all **x**, if **S(x)** then **P(x)** implies there is **x** such that **S(x)** and **P(x)**, is true does not lead to the consequent, which is, in this case, there is **x** such that **S(x)**, being true because of what you can see on (Pinheiro, 2016).

References

Philosophy Index, 2016 Tautology. http://www.philosophy-index.com/logic/terms/tautology.php

Pinheiro, M. R. (2016). http://itshouldallbeaboutlogic.blogspot.com.au/2016/05/existential-import-other-sources.html

What we paste below is (Pinheiro, 2016), which has just appeared as a reference here, in this very book.

Existential Import, Other Sources

Because I thought I had seen what Professor Corcoran had on his work somewhere else, perhaps in another language, and because I have been called now to various meetings about the Existential Import, I actually decided to do more research on the theme.

Britannica (2016) seems to think that Existential Import is something slightly different. See yourself:

> Existential import, in syllogistic, the logical implication by a universal proposition (i.e., a proposition of the form *All S is P* or *No S is P*) of the corresponding particular statement (i.e., *Some S is P* or *Some S is not P*, respectively). The validity of some syllogistic figures (see syllogism) depends on whether universal statements are interpreted as having existential import.

To correspond, from **Purely Human Language**, could probably be translated into **iff** from **Classical Logic**, but there is also a sense of this sigmatoid, correspond, that is slightly different: We could simply mean interchangeable when we say correspond, not necessarily something that implies the other and is implied by it. When we use the **material implication**, we have to follow the rules of Classical Logic, which are not necessarily attached to human meaning.

We say $5 + 3 = 8$ implies $5 = 8 - 3$, and we know that $5 = 8 - 3$ **does imply** $5 + 3 = 8$. In this way, it has to be true that $5 + 3 = 8$ iff $5 = 8 - 3$, since we can go both ways with the arrow and we will be right.

When we say correspond, however, we may mean something else. We could say that $8 - 5 = 3$ **corresponds to** $10 - 2 = 8$ because we are thinking of the operation and how all is numbers, etc. Notwithstanding, **it is not true that $8 - 5 = 3$ implies $10 - 2 = 8$**, and **it is also not true that $10 - 2 = 8$ implies $8 - 5 = 3$**.

Corresponding is something loser than **an iff**, which is what we keep on saying (please read Words for Science).

Saying that **All** corresponds to **Some** can be something related to simply having **Some S is P** in place of **All S is P** in a statement and looking at the shape of things, like perhaps that is all the same for me because I am an Engineer, for instance.

We don't see Britannica tying the sigmatoid **correspond** to **iff**, so that what they are saying seems to be completely different from what Professor Corcoran said. Yet, they use exactly the same expression Professor Corcoran used in the extract we here mentioned.

See this one (W. W., 1998):

Existential Import

A statement has existential import when its truth depends on evidence for the existence of things in a certain category—in the case of categorical propositions, the existence of things in the categories signified by its subject and predicate terms.

Existential Import and Categorical Propositions
Consider the following propositions:
All unicorns have horns.
No perpetual motion machine has been patented.
Both are true even though there are no unicorns or perpetual motion machines.
Thus, these statements lack existential import.
Many modern logicians hold that existential import is a function of a statement's logical form. According to this view, universal categorical statements in general do not have existential import.
Statements that are particular in nature, however, do have existential import. To say that some S are P, or that some S are not P, is to imply the existence of Ss: If there are no Ss,   then   both   statements   are   false.

We now have actually just read arguments against the Professor Corcoran's Existential Import, since what we have just read tells us that sometimes it is possible to have the existential import, sometimes it is not. That is exactly what we said ourselves in our blog post.
See (Hill, 2016) now:

EXISTENTIAL IMPORT

A proposition is said to have existential import if the truth of the proposition requires a belief in the existence of members of the subject class. I and O propositions have existential import; they assert that the classes designated by their subject terms are not empty. But in Aristotelian logic, I and O propositions follow validly from A and E propositions by sub-alternation. As a result, Aristotelian Logic requires A and E propositions to have existential import, because a proposition with existential import cannot be derived from a proposition without existential import. Thus, for Aristotle, if we believe it is that "All unicorns have one horn" then we are committed to believing that unicorns exist. For the modern logician or mathematician, this is an unacceptable result because modern mathematics and logic often deal with empty or null sets or with imaginary objects. A modern mathematician might, for example, wish to make a true claim about all irrational prime numbers. Since there are no irrational prime numbers, Aristotle would say that any claim about them is necessarily false.

By all the reasons above, I now believe that Professor Corcoran is wrong about the own definition of what an Existential Import is even having been born speaking a language that is not English.

What he refers to, the assertion he most plays with, which is (Corcoran & Masoud, 2014)

$$\forall x \, [S(x) \rightarrow P(x)] \text{ implies } \exists x \, [S(x) \,\&\, P(x)] \text{ iff } \exists x \, S(x) \text{ is logically true.}$$

is perhaps at most one possible instance of the Existential Import. He calls this the Existential Import Equivalence.

So, here, we criticize the very kind and generous, also very nice, Professor Corcoran, who is kind, generous, and nice so far with us, as for all we know, for the equivocated application of such a term, *Existential Import*, like we believe he should have been more careful when writing about such an element. It seems clear that the definition of what an Existential Import is is much broader than what he makes us believe, which is that it is all limited to that particular assertion.

We have previously criticized the own assertion on a previous post, and we would like to make the reader notice that the source that we presented here, which brought the own definition of the term Existential Import, seems to agree with us ipsis litteris. See:

> For the modern logician or mathematician, this is an unacceptable result because modern mathematics and logic often deal with empty or null sets or with imaginary objects. A modern mathematician might, for example, wish to make a true claim about all irrational prime numbers. Since there are no irrational prime numbers, Aristotle would say that any claim about them is necessarily false.

Now, the same source that gave us the extract above has mistakes in its text. See:

> As a result, Aristotelian Logic requires A and E propositions to have existential import, because a proposition with existential import cannot be derived from a proposition without existential import. Thus, for Aristotle, if we believe it is that "All unicorns have one horn" then we are committed to believing that unicorns exist.

They say that *a proposition with existential import cannot be derived from a proposition without existential import*. The problem is their own conclusion: That is what is wrong. They try to exemplify what they think Aristotle said, but their reasoning DOES NOT derive: *All unicorns have one horn* (generic statement that needs us to check each unicorn, and therefore that has an existential import. IF unicorns exist. IF unicorns do not exist, however, anything that is said about them could be true) does not necessarily have an existential import. We are definitely not committed to believing that they exist just because someone said that they all have one horn, not according to what they let us know this far about Aristotle. *None of the presented premises has a mandatory existential import*, and therefore none of them fall inside of the set of possible examples of what they claim to be the Aristotelian Logic (*a proposition with existential import cannot be derived from a proposition without existential import*).

They do mention however, in another piece, the following:

A modern mathematician might, for example, wish to make a true claim about all irrational prime numbers. Since there are no irrational prime numbers, Aristotle would say that any claim about them is necessarily false.

Void would therefore imply falsity, not truth. With this, their previous premise, *All unicorns have one horn*, would demand that we had unicorns if we had declared it in good faith (willing that it were logically true). In this case, it would have an existential import. The problem now is what they called Aristotelian Logic: If such a logic would really have this condition, that void would imply falsity. Upon checking a few sources, such as (Groarke, 2016), we found only reassurance. Aristotle seems to agree with us: That if there are no unicorns, then nothing can be said as to the truth-value of the mentioned premise, and therefore it is NOT the case that it is false if unicorns do not exist, but that its truth-value is not something we can determine instead.

See a relevant extract from (Groarke, 2016):

> He writes, "no one knows the nature of what does not exist—[we] can know the meaning of the phrase or name 'goat-stag' but not what the essential nature of a goat-stag is." (II.7.92b6-8, Mure.) Because we cannot know what the essential nature of a goat-stag is—indeed, it has no essential nature—we cannot provide a proper definition of a goat-stag. So the study of goat-stags (or unicorns) is not open to scientific investigation.

I feel that we are not as careful as we should in Science, and this is for very long, as you can see in all that I have been writing during all these years of martyrdom.

References

Pinheiro, M. R. (2016). Existential Import, Other Sources. Retrieved 19 December 2016 from http://itshouldallbeaboutlogic.blogspot.com.au/2016/05/existential-import-other-sources.html

Encyclopaedia Britannica, Inc. (2016). Existential Import. Retrieved 19 December 2016 from https://www.britannica.com/topic/existential-import

W. W. Norton & Co. (1998). Categorical Propositions. Retrieved 19 December 2016 from http://www.wwnorton.com/college/phil/logic3/ch8/import.htm

Hill, H. H. (2016). Existential Import. Retrieved 19 December 2016 from http://cstl-cla.semo.edu/hhill/pl120/notes/existential%20import.htm

Corcoran, J., & Masoud, H. (2014). Existential Import Today: New Metatheorems; Historical, Philosophical, and Pedagogical Misconceptions. Retrieved 19 December 2016 from http://www.tandfonline.com/doi/full/10.1080/01445340.2014.952947

Groarke, L. F. (2016). Aristotle: Logic. Retrieved 20 December 2016 from http://www.iep.utm.edu/aris-log/#H7

When we criticized the own assertion, the previous post as we said, we wrote the text we see now.

Corcoran and the Existential Import

I just had contact with Professor Corcoran through the Internet. He then had a link to part of his work. The link led to (Informa, 2016).

This part is about The Existential Import. I am a very curious person and I love exotic names and different things, especially if they refer to Logic. I then spied (my first language is not English and I think I was given another name for this assertion when I learned it):

$$\forall x\ [S(x) \rightarrow P(x)]\ \text{implies}\ \exists x\ [S(x)\ \&\ P(x)]\ \text{iff}\ \exists x\ S(x)\ \text{is logically true.}$$

[.. 01 ..] [.... .... 02 ....]

We notice that there is an *iff* inside of this assertion. In this case, we know that there are at least two assertions inside of it: One in the direction of the *if* and another in the direction of ***and only if.***

01 tells us that for all *x* it is true that if *x* satisfies the property *S*, then *x* satisfies the property *P*. All that this means is that if there is an *x* that satisfies the property *S*, so say *S = is a car*, then *x* will also satisfy the property *P*, so say *P = is black*.

We may however be in a parking lot where there are no cars, so say it is completely empty. In this case, it may be true that all that satisfies the property *S* will also satisfies the property *P*, but, given that we have absolutely no cars, nothing satisfies the property *S* and therefore nothing will satisfy the property *P*, and this will happen AT LEAST in the universe of that special parking lot we chose to hang on to for the purposes of this logical drill.

If we get the situation we just described, we will have a **1 -> 0** in Classical Logic, and therefore a material implication that is not true. From that, all follows, so that **02** will give us a true assertion. On the other hand, and going the other way around, that is, <=, if we assume that there is *x* such that *Sx*, we are not guaranteeing that there is *x* such that *Px*, so that the and, that is, the consequent of **01**, is not guaranteed, and therefore might be false. Notwithstanding, if it is false, then the antecedent might be true and we will have a false result and therefore a false implication, what then makes the entire thing false.

The problem then becomes the second part of what comes before the *iff.* The first part, which is the one we have just spoken about, would then imply that there is/are such an/such *x/xs*, basically, that is, that there is an *x* that satisfies both *P* and *S*, and therefore we have a black car in that parking lot. Notwithstanding, we have just said that the parking lot is empty, so that unless we want to enter the field of Metaphysics, that is not possible, so that one DOES not imply the other at all. Interesting is then that if we use the *iff* then it may work because if we were told that there is an *x* such that *x* satisfies the property being a car, then we know that that will imply this car is black and because the second part of **01** will be true, we have a **1 -> 1** in Classical Logic and that is always true, so that the entire assertion is true. That would be the, let's say, **if** part of it being completely verified (not the other **f** or the **and only if**).

We now do the ***and only if,*** and I will now opt either for diplomacy or wisdom: It is diplomacy if what I will call Brazilian School is correct, and it is wisdom if what I will call American School is correct. Considering the Brazilian School, we would have a double arrow each, and every, time we have an *iff,* so that the logic of proving should be the same in both directions. In this case, we would have proven the normal arrow direction or <- with the previous paragraphs and it would now be missing proving the direction ->, which cannot be proven inside of the Brazilian School, since we may not have *x* in reality, such as in the example we invented, of the empty parking lot.

Following the American School, ***If and only if*** would mean what the expression literally means, so that it is definitely the truth: ***If and only if*** we have the second part, what comes after the *iff,* as something that is logically true, will we have what precedes the *iff* being true.

The difference is then shocking. What the Brazilian School does would definitely agree with what Leithold and Bartle, both seen as Americans by Brazilian people, do. They are both dead. That means that there is no such a thing as the Existential Import if the Existential Import is what we see above (that single assertion). The American School would however sign under Corcoran's work if that is his own assertion and we are right in assuming that this assertion is what he has called Existential Import.

Since I love enriching things, I would kindly suggest that we embrace both points of view in an orderly manner and in a manner of being more logical than we have been: Let the double arrow have the Brazilian meaning for the *iff* and the American School *iff* not be represented by any symbol that there is, so that, from now onward, let's say people accept my suggestion, we will use *iff* to literally mean *if and only if* and *double arrow* to mean *double Classical Logic implication*, two directions. We could then add a **b** to mean *Brazilian School iff* whenever what is meant is a double arrow, so say *iffb*.

Now that you have believed all that, I will get you once more: The truth is that if what comes before the *iff* is a falsehood, which is obviously the case, as explained before, and what comes after the *iff* is a truth, we have a *0 1* situation in the -> direction, so that the implication should be true (the only way it would be false is *1 0*. The problem now is answering the question: What actually appears   after   the *iff* or   as   a   consequent?

In Logic, it should be the entire thing, so that it should be *there is x such that x is a car is logically true*. In this case, let's call this sentence *W*, so that *W = there is x such that x is a car is logically true*. Now *W* may be true or false. If *W* is false, then we have a *0 0* situation and therefore something true. If *W* is true, then we have a *0 1* situation and therefore something also true. Oh, now perhaps both schools coincide in what they intend, what is expected. Notwithstanding, the problem persists: If we use the normal sense of implies and apply to the *Brazilian School iff*, we have a problem with normal language, since we now can say that one thing does not imply the other in the -> direction. We call this the difference between Purely Human Language and Machine-friendly language in our work.

Now, I am going to change the context and make it all swing in your head again (I am actually a fan of Contextualism): In Logic, we can certainly (and we should) go only with the numerical values of the tokens we have, so that, to adequately analyse this assertion, we should actually split it into small bites, basically, and then assign Classical Logic values to each bite, finally attaining an overall result from that effort. With this, we would be getting a table. See:

A: ∀x[S(x)->P(x)]	B: ∃x[S(x)&P(x)]	C: A=>B
1	1	1
1	0	0
0	1	1
0	0	1

D: V(∃xS(x))=1	C	C => D	D => C
1	1	1	1
1	0	1	0
0	1	0	1
0	0	1	1

What we see now is that it is possible to get both directions failing: That will happen if **V(A->B)=1** and **V(D)=0**, for instance. In that case, every car in the parking lot will be black does imply that there are things in the parking lot that are cars and black. At the same time, all things in the parking lot are not cars. And it will also happen that **V(∃xS(x))=1** and **V(A->B)=0**. In this case, there are cars in the parking lot, but even though if there are cars they will be black, that does not imply that there are elements there that are cars and are black. Whilst this one does not seem too plausible, the previous objection is pretty much OK, the one we presented right before this last one.

Now, we are back to thinking that there must be Brazilian and American School of Classical Logic, just because of the difference in how we analyse the *iff*.

The interesting thing is that Professor Corcoran can be justified in all he states, assuming he states that the main assertion of this page is true because it is just a matter of how you write things. He wanted to make it all fit one logical sentence, but not only he has used mixed symbols, since he uses iff instead of double arrow, for instance, but he is also allowing for double meaning. We could easily rewrite his assertion to make it all be true: It is just that we would then have a couple of assertions rather than just one, this in terms of what the eyes can see. We could write things like this: Assume that it is true that there are elements that obey the property **S**. We then have **∀x[S(x)->P(x)]-> ∃x[S(x)&P(x)]**.

Logic is very much about how we write things, I reckon. I always aim at maximum clarity of exposure, so that I would recommend never putting it all in a single assertion in a case like this. Even if you write the **V** for evaluation instead of saying logically true, you still get a possible branch of interpreting that will match mine here.

References

Pinheiro, M. R. (2016). Corcoran and the Existential Import. Retrieved 20 Dec 2016 from http://itshouldallbeaboutlogic.blogspot.com.au/2016/05/corcoran-and-existential-import.html

Informa UK Limited. (2016). Taylor & Francis Online. Retrieved 20 Dec 2016 from http://www.tandfonline.com/doi/full/10.1080/01445340.2014.952947

This was a post with WordPress (Pinheiro, 2016).

The idea that follows is extraordinary: Inductive numbers. We could have numbers that are part of induction and call them inductive because of that or numbers that command the induction and call them inductive because of that. Because induction is based on the natural numbers, we immediately wonder why we would need an inductive number.

The mathematical induction has inspired lots of philosophers however. We discussed The Sorites Paradox for even a few millennia before we solved it (2000).

► JOHN CORCORAN. *Omega arguments in the 1931 Gödel formalism.*
Philosophy, University at Buffalo, Buffalo, NY 14260-4150, USA
E-mail: corcoran@buffalo.edu

A (premise-conclusion) argument is an ordered pair consisting of a set of sentences, its premise-set, and a single sentence, its conclusion. In a (logically) valid argument, its conclusion is a (logical) consequence of its premise-set. Informally, an omega argument has as its conclusion a universal sentence 'every number has (property) P', whose numeral instances '0 has P', '1 has P', '2 has P', and so on, make up the premise-set.

This paper treats arguments in the formalism of the 1931 Gödel incompleteness paper where the numbers are exhaustively denoted by the numerals '0', 's0', 'ss0', etc., 's' being the symbol for 'the successor of' . In this formalism, PMI, the Principle of Mathematical Induction, is (logically) equivalent to the sentence (translated) 'every property that belongs to every inductive number belongs to every number', called Eliminative Mathematical Induction EMI. As usual, 'is inductive' abbreviates ' has every property belonging to zero and to the successor of every number to which it belongs'.

I doubt I will ever grasp the meaning of the expression that appears in the above extract: Inductive number.

Such things sound like ghosts to me: Others say they saw, others swear, but I myself always think they are a bit crazy each, and every, time they mention their experiences.

An inductive number is perhaps a concept that appears as a consequence of thinking that the world is made of chocolate (Pinheiro, 2016a) and that every property (Pinheiro, 2016b) applies to all in a generalized manner.

To the record, what we are talking about here is the sentence that appears in the end of the above extract: The principle of mathematical induction is (logically) equivalent to the sentence (translated) *every property that belongs to every inductive number belongs to every number.* It is also about another sentence that follows this one: As usual, *is inductive* abbreviates *has every property belonging to zero and to the successor of every number to which it belongs.*

I think I know that we use the natural numbers to appeal to the glories of the mathematical induction and those can start with 0 (some say they start with 1 instead).

It might be true that we can start with any natural and, if the basic step of the mathematical induction is satisfied (if it is valid for *n* then it is valid for *n+1* is a true implication in Classical Logic), then the property is true for every natural number.

That obviously never meant it is true for every number, for we have negative numbers, rational and irrational ones, for instance.

Whenever we say things in Science, we have to be extra careful because Science is an attempt to make our insights become eternal, available for use for the entire human kind as absolute truths.

That is why I wrote Words for Science: If words are irrelevant to the message we intend to pass to someone else during most of the time, in Science they are of fundamental importance because if words that are slightly equivocated are used, the consequences are catastrophic (thousands of years of priceless resources applied in research on that topic, for instance).

No, it is definitely not true for every number, Professor Corcoran.

References

Pinheiro, M. R. (2016). Inductive Numbers: Prof. Corcoran. Retrieved 20 Dec 2016 from https://drmarciapinheiro.wordpress.com/2016/12/03/inductive-numbers-prof-corcoran/

Pinheiro, M. R. (2016a). Made of Chocolate Too. Retrieved 20 Dec 2016 from https://drmarciapinheiro.wordpress.com/2016/12/02/made-of-chocolate-too/

Pinheiro, M. R. (2016b). Every Property. Retrieved 20 Dec 2016 from https://drmarciapinheiro.wordpress.com/2016/12/02/every-property/

Pinheiro, M. R. (2015). Words for Science. *Indian Journal of Applied Research, 5*(5), 19–22. Retrieved 19 December 2016 from https://www.worldwidejournals.com/ijar/articles.php?val=NjQ0MQ==&b1=853&k=214

And now we have Every Proposition, the post we have just mentioned:

EVERY PROPERTY

When I listen to that song, from Cadbury (Alentsa, 2008), wouldn't it be nice if the world were… wouldn't it be nice, I feel as if part of the song and propaganda choices have been my suggestion for some reason. Those who know me inside out, and those do not necessarily know me in that way because I want, it is just, let's say, a criminally imposed condition of existence in the last decade and a half perhaps, would know that it is quite possible that whatever I imagine ends up being written by the hands of someone else, very unfortunately, and therefore the money and fame involved also ends up in the hands and accounts of someone else. Anyway, that is irrelevant for what I was trying to say here. I will go back to the main point.

Wouldn't it be nice if we could always generalize things in Mathematics? Perhaps everywhere, so say in Geography, History, Archaeology, and all else?

Basically, that saves our resources, is it not?

An archaeologist sees a stone and finds out it came from the tomb of Christ. Around that stone there is another stone, and he then says: Trivially, since this stone is close to the previous stone, and I have proven that the first stone came from Christ's tomb, this one also did. Therefore I don't need to go through the same effort here.

Wouldn't it be nice? Made of chocolate?

I found this amazing piece in Professor Corcoran's work (Pinheiro, 2016):

> Every property that belongs to every number whose predecessors all have it belongs to every number.
>
> In order for a property to belong to every number it is sufficient for it to belong to every number whose predecessors all have it.
>
> In order for a property to belong to every number whose predecessors all have it, it is necessary for it to belong to every number.
>
> CIP in symbols: $\forall P[(\forall x(\forall y(y < x \rightarrow Py) \rightarrow Px) \rightarrow \forall xPx]$

Oh, well, counter-examples are very easy to be found, Professor Corcoran: Every number below, or equal to, 1000 is less than, or equal to, 1000, but all numbers above 1000 do not have that property, of being below, or equal to, 1000.

The extract, instead of being made of chocolate, contains a piece of very bitter strawberry.

Not really.

Sorry to happily disagree \-/.

If this is CIP, then down with it, Professor Corcoran, down with it.

CPI is a meaningful acronym in Brazil, however, something like corruption commission.

This extract came from a page, and this page can be accessed through (Pinheiro, 2016).

Observation

In response to this post, a fellow called Peter wrote:

Your example doesn't work. Let Pz be $z \leq 1000$ and instantiate x as 1001 Then 'for all y ($y \leq 1001$ -> $y < 1000$)' is true but '$1001 \leq 1000$' is false so the antecedent for the implication is false.

In response to Peter's observation, I come up with the following steps: take P to be less than 1001. Take x to be 1000. Now we have:

(For all y ($y < 1000$ -> y less than 1001)->1000 is less than 1001)-> For all x, Px.

Because x = 2000 is a counter-example to the right side of the implication, it is not true.

References

Alentsa. (2008). Funny Commercial – Cadbury Chocolate – "Wouldn't It Be Nice?" Retrieved 21 Dec 2016 from https://www.youtube.com/watch?v=3KMfG-sox00

Pinheiro. (2016). Teaching Course-of-values Induction. Retrieved 21 Dec 2016 from https://drmarciapinheiro.files.wordpress.com/2016/12/teaching-course-of-values-induction.pdf

On the blog post that follows (Pinheiro, 2016), you will see the move of doing things in the other around (instead of going backwards, you go forwards, basically). See how interesting this all is, please.

MADE OF CHOCOLATE TOO

If it is valid for what is behind, then it is valid for what is ahead: Another one that matches the Cadbury System, as I shall call it (you wish for eating, eat it: It will then disappear).

JOHN CORCORAN. *Mathematical induction and specific-case semantic omega properties.* Department of Philosophy. University at Buffalo. State University of New York. Buffalo. NY 14260. USA.
E-mail: corcoran@acsu.buffalo.edu.
   Although the axiomatizations of second-order number theory preferred by mathematicians take as primitive entities (besides the universe of natural numbers beginning with zero or with one) the initial number and the successor function, there are axiomatizations whose languages do not involve numerals. i.e.. terms composed of a symbol for the initial number preceded by zero or more symbols for the successor function.
   Taking the *property* of being initial (instead of the initial *number*) and taking the *relation* of [immediate] succession (instead of the successor *function*). the following property-universal proposition can serve as the principle of mathematical induction:

Every property that belongs to every initial number and to every number that succeeds a number to which it belongs also belongs to every number.

So, the initial number is 1000 again, just to copy them: They repeat, we repeat the logic involved.

Being greater than or equal to 1000 is valid for 1000 and for its successor as well as the successor of every successor. Therefore it is NOT valid for every number: Only valid for those above 1000 or equal to 1000.

Not made of chocolate either, Professor Corcoran, not at all.

Sorry again \-/.

Professor Corcoran, despite what I say about Jesus, I am sure he loves me.

God also does, despite my logical/scientific proof of His possible non-existence.

Let's please be one with Science and for Science.

Thanks for the invitation, Professor Corcoran, I really appreciate your humbleness and capability of asking us to criticize your work. I know that many would never do that, so that you are really spectacular just for that one.

Please feel invited to also criticize my work in the same way, and please do that with the same kindness and love for Science and for whoever is voluntarily involved with it. Please observe that I absolutely always pass the tokens to you first, as I did with Priest, Sever, and everyone else whose work I have criticized.

Shame I could not have the opportunity of putting my beautiful academic skills out there: Priest, Sever, yourself, etc., all had this opportunity, however, and that is why I can criticize your work (basically, you presented it properly, you produced as much as you wanted and you freely did that, you published in best vehicles, etc.). The reason for me not to be able to do that is the same reason that was the force behind every action of my relatives against me, and it is now one decade and a half suffering the worst world atrocities because of those: I couldn't do what they did in the same way they did. I could not be better either. I could only be something inferior to them, the cost involved being irrelevant for all of them. All that mattered was guaranteeing that I would not be a greater star, especially during the term of their lives.

References

Alentsa. (2008). Funny Commercial – Cadbury Chocolate – "Wouldn't It Be Nice?" Retrieved 21 Dec 2016 from https://www.youtube.com/watch?v=3KMfG-sox00

Pinheiro, M. R. (2016). Made of Chocolate Too. Retrieved 21 Dec 2016 from https://drmarciapinheiro.wordpress.com/2016/12/02/made-of-chocolate-too/

We find never-ending wealth in the work of Professor Corcoran. You will see, in the blog post that follows (Pinheiro, 2016), how interesting this thought, of positive and negative applied to the terms we use, that is, to the lingo of the logical systems, is.

There is another much less important fact about first-order and basic logic that is worth mentioning. For this we have to divide the logical concepts into *positive* and *negative*. Without going into the details, let me say that there are no surprises here. "Every", "Some", "Is", "And", "Or", "If", etc. are positive. "Not", "No", "Distinct", "Nor", etc. are negative. The result is that every contradictory first-order proposition involves at least one negative logical concept.

(Corcoran, 1998)

Here, Professor Corcoran talks about logical concepts, but mentions sigmatoids instead. A concept should involve more than a title.

Then he talks about splitting them into negative, and positive, and we have already seen that happening in the work of someone else in our blog with Blogger that deals with Logic, and that was Professor McHugh's.

We know that that cannot work for several reasons: the reasons we have already presented there, on that blog post where we mention the reasoning of Professor McHugh, and his positive and negative, and other reasons.

Professor Corcoran says that, with no surprise, we get that *every* is a positive concept. Language depends on its application, since that is how it started. We need a context. We say, every person you saw in that room is dead, and we have a very negative thing associated with the sigmatoid *every*. We say, every single person you spoke to on that day became a millionaire, and that is a really positive thing.

He then goes on to say that *no* is a negative thing, but, once more, see: No, you won't die. And then see: No, I don't love you. The former is positive, and the latter is negative, since the sigmatoid has no value in isolation in what comes to assessments of this type.

Every contradictory preposition in first-order logic involves at least one negative concept is what comes next. No, no, no, I won't die, but no, no, no, I can't live. This is contradictory, and does involve what he classified as a negative concept. Now, what about this one: Yes, yes, yes, I will die, but I will live forever. All the things he called concepts are non-negative, and, even so, it is contradictory.

References

Corcoran, J. (1998). Second-order Logic. Logic, Meaning, and Computation: Essays in Memory of Alonzo Church (C. Anderson et al., editors), Kluwer.

Pinheiro, M. R. (2016). Positive and Negative in Logical Systems. Retrieved 21 Dec 2016 from https://drmarciapinheiro.wordpress.com/2016/09/12/positive-and-negative-in-logical-systems/

The contents below came from (Pinheiro, 2016), and this blog post was about implication, which is perhaps the most important concept of all in Logic and logical deductions: We realistically need to know what follows from what and why.

CORCORAN: MATERIAL IMPLICATION x LOGICAL IMPLICATION

Professor Corcoran in The Founding of Logic, Modern Interpretations of Aristotle's Logic (Corcoran, 2016), states that:

"To see that not every truth-preserving process is consequence-conservative it is sufficient to consider the rule of mathematical induction which, for example, when applied to the two propositions 'zero is even' and 'every natural number which is the successor of an even natural number is even' results in 'every natural number is even.' This resultant is materially implied by the given premise-set since the second premise is false and, of course, every proposition is materially implied by every set of propositions having a false member. On the other hand, the resultant is not logically implied by the given premise-set. To see this use the method of counterarguments: 'zero is integral' and 'every real number which is the successor of an integral real number is integral' are both true whereas 'every real number is integral' is false, of course; one-half is not integral, for example."

My remarks are as follows:

We all know that (0 and 0) does not give us 1 in Classical Logic, right? We also know that (1 and 0) does not give us 1 in CL.

Slide 7 of (Leifer, 2009)

Let's assume, for the sake of the argument, that 0 is an even number, since plenty of people defend that on the Internet. In this case, you have a 1. Even so, the successor of an even is not even, so that the second premise is wrong and you then have to agree that you have a 0 there. As said before, (1 and 0) will give you a zero in CL. Professor Corcoran agrees with all this, for he is saying that it is not logically implied.

zero is even and *every natural number which is the successor of an even natural number is even* results in *every natural number is even.*

Here you must notice that this is not a Classical Logic inference because we do not have a few essential premises, such as every natural number is zero or is an even number or is the successor of an even number.

If we were to code that, we would get perhaps A = zero is even, B = every natural number which is the successor of an even natural number is even, C = set of the even numbers, D = every natural number is even, and s(x) = successor of the natural number x. That would result in $(0 \in C \land x \in N \land x = s(y), y \in C) \Rightarrow (x \in N \Rightarrow x \in C)$ or $A \land B \Rightarrow D$.

Since 0 is an even number, this according to your assumptions, v(A)=1. B is obviously false, so that v(B)=0. With this, v(A ∧ B)=0. V(D)=0 is then as acceptable as v(D)=1, since, when the antecedent is 0 in Classical Logic, all can be implied. If so, we cannot really say that D is true or that D is false, what then matches our information: We do not have enough premises to hang on to. In this case, it is not sound to say that A ∧ B results in D or that

> *zero is even* and *every natural number which is the successor of an even natural number is even* results in *every natural number is even.*

Professor John then talks about material implication and the fact that any falsehood, of any premise, would give us anything in exchange, but, I think that would be a reference to something like the Explosion Law and therefore it only happens if you have an implication and a false antecedent, like we see in (Weisstein , 1999).

In this case, when the antecedent is false, the consequent is irrelevant for the result, which is always true. I think that is what Professor Corcoran was referring to here, but it (maybe) ended up containing a bit of inaccuracy.

Here we seem to have a conjunction, and therefore there is no explosion. We always get a zero as a result instead, not one, if you want it to be logically valid. As a result, there is no material inference, only logical: v(A ∧ B ⇒ D)=1 if we assume that v(A)=1 and v(B)=0, like regardless of v(D). Notwithstanding, it is clearly not a material implication because when we put the suggested words there, we do not get soundness.

Having said that, I directed myself to (Encyclopaedia, 1998) and found out that some people say that Material Implication is the same as Logical Implication. In this case, the implication mentioned by Professor Corcoran is actually always materially true.

See:
> In most systems of formal logic, a broader relationship called material implication is employed, which is read "If *A*, then *B*," and is denoted by $A \supset B$ or $A \rightarrow B$. The truth or falsity of the compound proposition $A \supset B$ depends not on any relationship between the meanings of the propositions but only on the truth-values of *A* and *B*; $A \supset B$ is false when *A* is true and *B* is false, and it is true in all other cases. Equivalently, $A \supset B$ is often defined as $\sim(A \cdot \sim B)$ or as $\sim A \lor B$ (in which $\sim$ means "not," $\cdot$ means "and," and $\lor$ means "or"). This way of interpreting $\supset$ leads to the so-called paradoxes of material implication: "grass is red $\supset$ ice is cold" is a true proposition according to this definition of $\supset$.

What could then happen is that the implication would not be true in terms of real life, since it is missing elements for us to connect those things as we do, let's say, in a police investigation. We should then invent a new term for that, so say Natural Implication, since Logic is more than logic, which connects to logical systems.

Clarke (1996) states that

> A large volume of research shows that humans reason poorly about conditional statements and that the formal notion of material implication is difficult to learn.

Textbooks on Logic have used a variety of approaches to the introduction and justification of a truth-functional definition of material implication.

and that

Most commonly, material implication is defined by truth table or some verbal equivalent such as "X-->Y is always true if X is false and also if Y is true" [HILB50 p4] or "A conditional sentence is false if the antecedent is true and the consequent is false; otherwise it is true" [SUPP57 p6].

We are then relieved because the term seems to have brought confusion to the writing and understanding of logicians.

(Planetmath.org, 2016) defines Logical Implication and it then looks like both concepts will coincide in the case we here mention: Both Logical Implication and Material Implication.

I did not feel the necessity of doing the Brazilian and American thing here, just of creating a new term, which is then Natural Implication, to denote what we feel in terms of real life and the Material Implication from Mathematics/Classical Logic.

References

Pinheiro, M. R. (2016). Corcoran: Material Implication x Logical Implication. Retrieved 22 Dec 2016 from http://itshouldallbeaboutlogic.blogspot.com.au/2016_05_01_archive.html

Corcoran, J. (2016). The Founding of Logic Modern Interpretations of Aristotle's Logic. Retrieved 22 Dec 2016 from https://www.academia.edu/s/0cbb00f5e7/the-founding-of-logic-modern-interpretations-of-aristotles-logic

Leifer, M. (2009). Quantum Logic. Retrieved 23 Dec 2016 from http://www.slideshare.net/mleifer/quantum-logic

Weisstein, Eric W. (1999). Implies. From *Math World*-A Wolfram Web Resource. http://mathworld.wolfram.com/Implies.html

Encyclopaedia Britannica, Inc. (1998). Implication. Retrieved 25 Dec 2016 from https://www.britannica.com/topic/implication#ref289368

Clarke, M. C. (1996). A Comparison of Techniques for Introducing Material Implication. Retrieved 25 Dec 2016 from http://www.cs.cornell.edu/Info/People/gries/symposium/clarke.htm

PlanetMath. (2016). Logical Implication. Retrieved 29 May 2016 from http://planetmath.org/logicalimplication

The text you are about to read appeared on our blog post with WordPress (Pinheiro, 2016).

MAZUR, ENTITY, AND ORDER: PROF. CORCORAN NOTICED

We discuss the following semantic question—apparently about the denotation of the 9-character expression '5 mod 691' in traditional arithmetic—raised by Harvard mathematician Barry Mazur ([1] page 223).

"Is 5 mod 691 to be thought of as a symbol, or as a stand in for any number that has remainder 5 when divided by 691, or should we take the tack that it (i.e., "5 mod 691") is the equivalence class of all integers that are congruent to 5 mod 691?"

The expression '5 mod 691' is not in subject position in the true sentence '1387 is congruent to 5 mod 691'. If '5 mod 691'—as used in ordinary arithmetic literature—denotes an entity, one would expect the expression to occur as subject in its sentences, but we know of no sentences such as '5 mod 691 is congruent to 1387' using the problematic expression as subject.

Consider the following statement:

$$1387 \text{ is congruent to } 5 \text{ mod } 691$$

Could this statement be using a three-place relation

$$x \text{ is congruent to } y \text{ modulo } z$$

to interrelate three numbers: 1387, 5, and 691? If so, thinking that '5 modulo 691' is the object of the sentence—with '1387' as subject and 'is congruent to' as verb—is a *segmentation fallacy* related to *hypostasis* or *reification*: taking a non-denoting sentence fragment to be denoting an entity. Perhaps the "solution" to Mazur's problem is that there is no problem: perhaps he asked a pseudo-question. This article inquires how such issues can be objectively decided.

(Corcoran, 2016)

So, Doctor Mazur seems to have contradicted himself, first of all. Notice that he himself built a sentence where **5 mod (691)** was the subject (**5 mod (691) is the equivalence class of all integers that are congruent to 5 mod (691).**

In the paragraph that follows this sentence, he wrote: We know of no sentences such as 5 mod (691) is congruent to 1387 using the problematic expression as subject.

In Mathematics, *being congruent to* is a reflexive property, so that **x is congruent to y** does imply **y is congruent to x**. In this case, Doctor Mazur's discussion looks futile, as Professor Corcoran seems to have said.

As to the hypostasis, we should then say **5+2 is equal to 7**, never **7 is equal to 5+2**. It sounds unreasonable, quite sincerely. What about, which is a very similar assertion, **2 is the remainder of the division between 6 and 4**? We could never say things in this way: We should instead say that 6 divided by four gives us a remainder 2. It again does sound unreasonable.

I would subtract one point from each in the exam for weakness of argument and I would add one point to each for teaching us more about the differences between writing in normal language and writing in mathematical language.

As I keep on saying, we definitely enter a new world, the World of Mathematics, when we refer to the mathematical elements. Our world is way bigger, way more inclusive, and way more confused, especially in terms of communication.

In this case, however, when the world is adequately changed, we get that **Marta is sad because Tom and Mary, both, attacked her** is worse, in terms of token of discourse, than **Tom and Mary attacked Marta, and that is why she is upset**.

It seems again unreasonable to say that.

References

Pinheiro, M. R. (2016). Mazur, Entity, and Order: Prof. Corcoran Noticed. Retrieved 25 Dec 2016 from https://drmarciapinheiro.wordpress.com/2016/11/

Corcoran, J. (2016). Mazur's Semantic Problem. Retrieved 25 Dec 2016 from http://www.academia.edu/28033462/Corcoran-Prentice_on_Mazur_s_semantic_problem

I had contact with Professor McHugh's work because of Professor Corcoran and his sessions with Academia.edu. Someday Professor McHugh wrote something inside of the session that caught my attention, and we then started conversing.

Professor Christopher McHugh spoke about refined concepts, very refined concepts at least sometimes, as we can see in what comes in the fourth paragraph after this one.

Probability, according to him, would be an epistemic concept because the more we know about the context, the better we can determine it.

In the past, I would see several researchers asking for examples of what was being said, like for instances of application of whatever theory was being discussed, and I used to think to myself that such were not really necessary, that it all sounded obvious.

Nowadays, I more than support examples each, and every, time we discuss things, especially in Philosophy. I think it is missing a practical example in what you will see after this paragraph for us to really believe what is being said by Professor McHugh. Perhaps for lacking the actual situation, which he probably had in his Inner Reality as he wrote the text that follows, we cannot really see anything different from what we saw and later on stated in simple words.

PROBABILITY AND ENTAILMENT

Again, in Entailment II, an open session from Academia.edu,

Professor McHugh stated:

There are no entailments (broadly speaking) in matters of probability. We can, of course, speak of "entailments" within a specified *subjective* frame of reference for some probability calculation, but we can go no further than that. For example, if (given some set of data) someone correctly determines that the probability of some event is .5, we can say "a probability of .5 entails that the probability is not 1". However, outside the boundary of that computational situation, another person exposed to better data can say "the probability of the event is 1". Assuming right math and right method, both calculations are correct, given the data available to each, and this is not a contradiction. That's because probability is an epistemic concept, not an ontological or aletheic one.

Reply:

Probability is an epistemic concept, nor an ontological or aletheic one sounded like too much, Professor McHugh.

We assume that what you mean here is that we get to know probability from knowing the events, so that the more we know the events, the more we know about probability, this considering your practical example, contained in your text.

What would an ontological concept be, however?

Aletheic seems to mean (Farlex, 2003):
ALETHIC

Related to alethic: (Farlex, 2003a)
ALETHIC

(əˈliːθɪk)
adj
(Logic) *logic*
a. of or relating to such philosophical concepts as truth, necessity, possibility, contingency, etc
b. designating the branch of modal logic that deals with the formalization of these concepts
[C20: from Greek *alētheia* truth]

In this case, you mean that probability does not relate to concepts such as truth, necessity, possibility, etc.

It was now you who contradicted yourself then, Professor McHugh: If it does not have to do with truth, how come from being closer to it we get more accurate results? Like Jesus would put it: You said it. It was your own example.

On the other hand, how come probability would not have to do with possibilities? I find that a bit out of earth, like I thought it was all about possibilities instead. As Graham Priest would perhaps put it, there is the actual and the possibly realizable world.

As for ontological probability, would Lotto not be one example? I am not so sure we can think of essence of this game, but, in case we can, would it not be probabilities?

As another point, you seem to contradict yourself again in the beginning: You seem to say that Mathematics has a very exact way of calculating probabilities if we have the right information. If so, why would we not be able to produce entailment, like using your own reasoning, from the beginning of the line of argumentation?

The least we can say is, for instance:
We have a 1/6 chance of rolling a dice once and getting the number 6 on the face that we can see from a certain angle entails that it is not guaranteed that we will get the number 6 next time we roll this dice if it is not biased.

And we can also say:
In a dice that has all faces marked with the number 6, the probability of getting a six on the face we can see from a certain angle is 100% entails that we will get the number 6 next time we roll this dice if it works as usual.

Both paragraphs we see above this one are talking about situations that won't be changed if we get more information in what we could call Normal World. Were it Chemistry, they say CNTP in Portuguese, that is, Normal Conditions of Temperature and Pressure. It is Mathematics, so let's say Normal World = all things as expected, so, let's say, there is no earthquake, tsunami, and things like that going on. The dice is not being shot by a laser machine. And so on.

That does make of our probability something absolute, so that it is not really an epistemic situation, as you hypothesized. All our exams in Mathematics will contain this sort of situation, by the way, Professor McHugh: Normal World situations. Otherwise, it would not be fair marking the objective answers of our students right or wrong, but everyone on earth seems to agree that Yes, We Can, so that it is fair and it is therefore absolute.

References

Pinheiro, M. R. (2016). Probability and Entailment. Retrieved 21 Dec 2016 from https://drmarciapinheiro.wordpress.com/2016/10/04/probability-and-entailment/

Farlex, Inc. (2003). Alethic. Retrieved 21 Dec 2016 from http://www.thefreedictionary.com/Alethic

Farlex, Inc. (2003a). Alethic + logic. Retrieved 21 Dec 2016 from http://www.thefreedictionary.com/alethic+logic

McHugh, C. (2016). Christopher McHugh. Retrieved 21 Dec 2016 from https://independentresearcher.academia.edu/ChristopherMcHugh

Professor McHugh also had incredibly inspiring writing. He said, for instance, that everything that was worthy involved Classical Logic.

See the following blog post from Blogger (Pinheiro, 2016):

Anything Worthy Reduces to Classical Logic

Given my passion for Classical Logic (CL), one could think that the title is an assertion that came from my hand, but it wasn't. Believe it or not, who wrote this was Professor McHugh inside of a session revolving around my draft, Entailment II, a draft that I put inside of Academia.edu for discussion (open to everyone).

The paragraph that I copy below is his.
I reply:

"Whatever makes the left true" is the fact of being part of the whole of "the facts". This is also what "makes the right true", too. Yes, this means "entailment". Despite your stated aim, here, we can (and must) use the word "true" or the equivalent thereof. Though it's "reserved for classical logic", any non-classical logic worthy of consideration necessarily reduces to classical logic. If any logicians happen to think otherwise, it's just because they're educated beyond their intelligence.

(Wow, that was an incredibly offensive remark, wasn't it?)

Well, Fuzzy Logic has been used in practice in several areas by human kind, so that perhaps Professor McHugh is referring to such a scope of utility when he classifies a logical system as worthy. It does have the Classical Logic values (true and false) plus the values in the middle (even an entire real interval of values). It does not reduces to CL, however: We can actually use their idea and produce inferences from **0.5 of truth** to the antecedent and **0.6 of truth** to the consequent, like a **0.5 -> 0.6** could be a proposition included in our system. In this case, there was no reference to true or false, just included or accepted in the system. Yet, the system must be worthy. Now, I must confess that they originally translated the result into true or false, so that Professor   McHugh   could   be   right   here.

OK. Would Paraconsistent Logic, not the ontological version, which we have discussed in (Pinheiro, 2016a), but the non-ontological one, be worthy? It seems that they even found an application in robotics for it. In this case, we have CL that accepts contradictions, like a CL where those do not lead to explosion or to all possible assertions in the system. We go and select which ones will be accepted upon conflict of a certain type instead. The result, which is the implication that is now palatable, may as well be said to have a situation of true.

Can I think of any other NCL system that is worth it? Hyde liked Supervaluationism. In it, we also had the translation of the multivalued system into a CL situation (true would be their supertrue and false would be their superfalse). Notwithstanding, we also had things in the middle that would not end up called true or false. They would be true in some evaluations and false in others, so that they would be the orphan situations, unnamed, or blurred or something else. In this way, we have just found a logical system that is considered worthy by at least one important modern logician to serve us as a counter-example: It does not reduce to true and false in the end.

References

Pinheiro, M. R. (2016). Anything Worthy Reduces to Classical Logic. Retrieved on 21 Dec 2016 from http://itshouldallbeaboutlogic.blogspot.com.au/2016/10/anything-worthy-reduces-to-classical.html

Pinheiro, M. R. (2016a). Ontological Paraconsistency Has a Place. Retrieved on 21 Dec 2016 from https://www.researchgate.net/publication/301556137_Ontological_Paraconsistency_Has_a_Place

The post below originated in my first contact with Professor McHugh's writings.

Possible Hydrology and Logic

(The format is all crime, and I am without much patience, so that I won't go HTML again to fix alignment and things like that. Sorry. I think the contents are already miraculous enough)

So, I was recently invited for some online sessions that involved discussing other people's pieces, what does please me: This thing of giving opinion, analysing, etc.

Interesting enough is the fact that the first person who invited me for one of those, Professor Corcoran, was very quickly identified by me as someone who had worked with Doctor Priest for reasons that not many people would believe.

I then got invited to discuss the work of Professor McHugh, like soon after that, and he said he got kicked out of Professor Corcoran's sessions before he invited me to talk about his work. I then did some research online on his name, and found out, like as for websites, that he worked with Hydrology, and he was therefore likely to be friends with Bradley Paul Neal, who is the person to cause all my disgraces in life, and at work since at most the end of 2001, but he was already causing a lot of crime to my end, and I was actually trying to denounce it all in a place where there is no protection to the most fundamental democratic rights of women, second in my life, as for all I was seeing before speaking to Trevor. It was all pretty disgusting, and I have been giving a lot of information about all that in my blogs recently because all the systems seem to be corrupt, and immoral, and I am left with very little. If at least I got my permanent, and non-virtual, academic position, I would not be talking about academic crimes anymore, but I feel is a lot of hostility instead of support, as if people are taking advantage of my situation almost all the time or all the time, so that I have to keep on doing what I am doing. What I think is that if there is any chance a non-bugged person is out there, and they read my blogs, they will be better equipped to try to help me, and, maybe because of the blogs, they will succeed. I think that otherwise only my mortal enemies, who illegally possessed, truth be said, my entire being since that end of 2001, will be equipped, and therefore I will never have a chance. It is only by miracle that I can write here, and publish, however, since I spent a long time without being able to do it, actually more than ten years (they hacked my sites, changed contents, and so on).

Anyway, I will now publish the text I received from Professor McHugh, who is actually, as for his current profile with Academia.edu, a Philosophy Professor, here because my remarks got deleted from the Academia.edu system, I told him that I would be doing this, and, in this way, there might be some use for me in what I will say.

A proposition P is positive iff it specifies something being the case about beings and their relations. ~P is a negative proposition iff P is a positive proposition. The positive propositions are not merely those without a negation sign, but those which say something about what is the case concerning being(s) and/or relation(s) and/or properties. [For example, "John went to school" is a positive proposition. "It is not the case that John went to school" is a negative proposition. The latter says nothing at all about what is the case concerning beings/relations/properties. "It is not the case that John went to school" says nothing about what John does, and also doesn't even commit us to there being a "John". It also doesn't say anything about whether there even is such a thing as a "school".] The totality of positive propositions determines the totality of negative propositions and vice versa. If you have one set, you have the other, perfectly. No negative proposition entails anything positive other than whatever is logically necessary. There are no contingent positive propositions entailed by any negative proposition or any collection of negative propositions short of the totality of negative propositions. However, if you have the totality, it entails all of the contingent positive propositions whatsoever, down to the last. If you remove any one negative proposition from the totality, the remaining set entails none of the contingent positive propositions at all. If you put back that last negative proposition, though, then all positive contingent propositions are entailed. What's particularly weird is that *any* negative proposition can be arbitrarily taken as the "last one", and that's a lot of power for one proposition. What's going on here? Nobody knows, as far as I know, so that's why I ask.

This thing of positive, and negative propositions, and splitting the universe of propositions into only those two, sounds a bit odd. That should be part of the Cartesian thinking, and therefore of Classical Logic, first of all, but Professor McHugh mentions Contingency, and this is something that appears in Priest's World, and therefore in Nonclassical Logic, NCL, instead.

A double negation, as the other fellow also said in the discussion, would be something complex to analyse according to the text we see, is it not? Is it positive or negative?

Then, it is said that the totality of positive propositions determines the totality of negative propositions, but, once more, double negation would tell us that it is not like that.

[For example, "John went to school" is a positive proposition. "It is not the case that John went to school" is a negative proposition. The latter says nothing at all about what is the case concerning beings/relations/properties. "It is not the case that John went to school" says nothing about what John does, and also doesn't even commit us to there being a "John". It also doesn't say anything about whether there even is such a thing as a "school".]

This extract could be OK: John went to school is a positive proposition. If you think that going to school is a bad thing, however, that could be a negative proposition, right? Notwithstanding, we think we understand what he wants to communicate here. It is not the case that John went to school would be a negative proposition for him. OK, so perhaps it has to do with the presence of the not or something. He says that the last example says nothing about what the case concerning beings/relations/properties is. I disagree: Something is being said about the relationship between

John and the school, for instance. He did not go to school, so that there might be an implication that means negligent or something. Not doing something is an actual action in my understanding, so that even choosing not to act is an action. In this case, it is saying something about what John does. He says that that does not commit us to there being a John. Notwithstanding, if we say that John did not go to school, how can we possibly know if there is no John? Well, we could then imagine that someone was like guessing names, said John, and someone else checked and there was no John in the punched cards list or something. In this case, that John we have mentioned does not exist, so that there is indeed no commitment to that. Now, if there is no school, could we be in the same situation? Yes, sure. There is no school where he lives, the person is from somewhere else, and asks if he went to the school. It is not the case that he went is a possible answer even if there is no school around there.

> No negative proposition entails anything positive other than whatever is logically necessary. There are no contingent positive propositions entailed by any negative proposition or any collection of negative propositions short of the totality of negative propositions. However, if you have the totality, it entails all of the contingent positive propositions whatsoever, down to the last. If you remove any one negative proposition from the totality, the remaining set entails none of the contingent positive propositions at all. If you put back that last negative proposition, though, then all positive contingent propositions are entailed. What's particularly weird is that *any* negative proposition can be arbitrarily taken as the "last one", and that's a lot of power for one proposition. What's going on here? Nobody knows, as far as I know, so that's why I ask.

Logically necessary is again an expression from NCL. It is a bit weird having this talk about positive, and negative, and then NCL, but here we go: Why would a negative proposition, like let's assume the best-case scenario, and the hypothesis that double negations are excluded, and there are rigid, and coherent, rules to determine negative propositions, entail only what is logically necessary? Why couldn't we have a negative proposition entailing something positive that is not logically necessary?

See an extract of my paper on Completeness:

> 1.1.5 It is also standard to define two notions of validity. The first is *semantic*. A valid inference is one that *preserves truth*, in a certain sense. Specifically, every interpretation (that is, crudely, a way of assigning truth values) that makes all the premises true makes the conclusion true. We use the metalinguistic symbol '$\models$' for this. What distinguishes different logics is the different notions of interpretation they employ.

Now, I was just caught with the things I never saw anyone else discussing, what is interesting: If we have void entails something, which would be the case with something we consider to be always true, such as the example I mention after this paragraph, that should mean that void plus something else, so say a lose proposition, as it is the proposal, will entail that same something. Notwithstanding, I have never seen anyone talking about this with so much clarity.

Say we say *Marcia is not beautiful, she is a victim*, and we then analyse *it is not the case that Marcia is beautiful* just to follow exactly what Doctor McHugh proposes. I don't need it to be true that *it will rain tomorrow*, so that is not a logical necessity for me. It could be that the meteorology department said the same, that it is *only possible* that it rains. In my world, the meteorology department, and whatever they say is considered to be the truth, so that that is always true. Given that that is always true, the contingency, we could have anything at all to the other side, so that that would always be entailed in my system. Usually it is more than a proposition that entails something, and we also need to know what logical system we are talking about. I have just written a paper to talk about Completeness, and, in this paper, one of the things that is discussed is precisely the definition of entailment. I think that Priest, and his fellows from NCL are currently using a definition of entailment of CL-type for all their systems when checking completeness (observe that it is all about true and false), first of all, but perhaps they should question that. We are probably missing the definition of entailment that Professor McHugh refers to here, and also the name of the logical system he is relying on. Is he creating his own? That is the impression I had. If so, it is missing properly introducing this system to us, I reckon. From what I just said, one understands that what follows is nonsense, like taking away positive, and negative propositions, and then making assertions regarding entailment.

The last extract to comment on:

> A proposition P is positive iff it specifies something being the case about beings and their relations. ~P is a negative proposition iff P is a positive proposition. The positive propositions are not merely those without a negation sign, but those which say something about what is the case concerning being(s) and/or relation(s) and/or properties.

Every proposition should specify something being the case about beings, and their relations, I reckon. Once more, even if we say we don't relate to anyone, and that is our proposition, we are still saying something about us, and everyone, all beings, and also about how we relate. We would probably need more examples to understand the point under analysis here.

References

Pinheiro, M. R. (2016). Possible Hydrology and Logic. Retrieved 22 Dec 2016 from http://itshouldallbeaboutlogic.blogspot.com.au/2016/09/hydrology-and-logic.html

Professor McHugh is the author of the most offensive writing I have ever seen in what regards Nonclassical Logic, the most outrageous. The following blog post appeared in my blog with WordPress.

In an open session with Academia.edu, which we have called Entailment II, Professor McHugh wrote:

> For any non-classical system that's actually a system (i.e. it isn't just abject nonsense called a "system"), it can easily be shown that whatever's there reduces to classical logic, and that it really *is* classical logic (and math, perhaps, which reduces to classical logic) applied to some area of inquiry not directly accounted for using *only* the rules

of classical logic simpliciter. Necessarily, though, this application preserves the rules of classical logic, and if doesn't, the system is incoherent, and is therefore a non-system. Or, to suit the people who disagree with me, perhaps it's both a system and a non-system, and we can be consoled that they necessarily agree with me precisely because they don't.

We reply:

This part, where you say that we can say that they agree with you precisely because they don't, Professor McHugh, was quite sui generis.

As for the systems that cannot escape the Goddess, Classical Logic, Graham Priest seems to have an entire encyclopaedia to disagree with you. Please read his book from 2001, the one I studied with him in 2000.

I don't remember very well, but I do think several did not reduce to False and True in the end. He definitely claims to have proven soundness and completeness for each one of them. As you know, I am now arguing that he didn't. That is in my last papers, including the one you currently study, which I called Entailment II so far.

What makes you think that a logical system with the values True, Indeterminate, and False, such as the one I mentioned in (Pinheiro, 2016) (I switched to here because I am again getting problems there with spaces and things like that), and that was Supervaluationism, would not make it to the end of the competition?

They do their Maths and it works, it seems.

Say you choose that I -> T is T, and it is only T -> I that is F, just to follow CL somehow, why do you think this would not work in the end? We would still have true, false and indeterminate as a result of assessment for several items.

Obs.: Perhaps, to understand better what is going on, refer to (Pinheiro, 2016).

References

Pinheiro, M. R. (2016). Anything Worthy Reduces to Classical Logic. Retrieved 22 Dec 2016 from http://itshouldallbeaboutlogic.blogspot.com.au/2016/10/anything-worthy-reduces-to-classical.html

Pinheiro, M. R. (2016). Nonclassical Systems that Are Actual Systems. Retrieved 22 Dec 2016 from https://wordpress.com/post/drmarciapinheiro.wordpress.com/1268

Professor Silvestru Sever Dragomir

Professor Silvestru Sever Dragomir

MSc, DipEd, PhD Maths, Romania

Chair in Mathematical Inequalities

Professor Dragomir is the Leader of Applied Mathematics Group in CES and also Honorary Professor in School of Computational and Applied Mathematics, University of the Witwatersrand ◪, Johannesburg, South Africa.

He is the Chair of the international Research Group in Mathematical Inequalities and Applications ◪ (RGMIA) and the Editor in Chief of the Australian Journal of Mathematical Analysis and Applications ◪ (AJMAA). He is a member of the editorial boards of more than 30 international journals. As confirmed by the American Mathematical Society database he is the author and editor of 20 books ◪ and more than 800 publications ◪. The work of Professor Dragomir is highly cited as shown by his Google Scholar ◪, Academic Research ◪ and Research Gate profiles ◪.

Contact details

📞 +61 3 9919 4437

✉ sever.dragomir@vu.edu.au

(Victoria University, 2016)

We met Professor Dragomir in Melbourne, Victoria, at the Victoria University of Technology, Footscray.

Almost by the time we left VUT, we checked the MathSciNet, which is one of the most important vehicle of dissemination of results in Mathematics, if not the most important, and Sever scored an 800 there, that is, he had about 800 papers listed.

Those papers did include his Research Report papers, what means unpublished ones, but it is still a lot of results and papers. That far, the champion in publications I had in my mind as champion in Mathematics was Poincare, with about 800 by the time of death.

I would have 800+ by 2010 if never attacked, what would be about half the age of Sever when I meet him, what would have made of me the most published researcher ever alive. I tend to think a man studied my potentiality with whatever they did to me in that 2002/3 at RMIT and he then decided that there was no way they would let a woman born in Brazil, heterosexual, and that attractive, when all researchers in history we have heard about from Mathematics and Logic are considered, succeed in achieving so much. Perhaps the most important point was that I was a person who had religious beliefs, but Priest and his group had tried to convince me, in that 2000, that I could not have those and still do research. I told Trevor Skinner, in that end of 2001, however, that it was precisely the opposite: Only those who were righteous, Common Biblical Core considered, could ever have meaningful results, correct ones.

By making sure I cannot work in Science in the way that I was doing that far, what means fully supported by institutions that are regarded as academic, First World ones, having to do no other thing, and not suffering heavy crime, especially extreme violation of human rights, they guaranteed that I would never get that score, since it is now about 15 years of maximum violation of human rights to my side, and I definitely do not have peace, resources, and anything else that I would need to have those results, publications, lectures, etc.

Nothing counts more in research than numbers, I reckon.

I have just heard about the death of Pete Burns. He had one song that was a hit, in my humblest, nothing else but a song.

We do care, watch (02091977BX, 2014), but he does not compare to someone like Freddie Mercury (Queen, 2008), who had more than 10 hit parades.

Not only we get that impression, that the person is realistically talented, just from hearing about the figures involved, so that we feel way more attracted to them, to watching what they presented or to read what they published, but we have an expectation before reading their writing or watching their presentation, which is that those pieces of basically art, since I do believe that everything that involves human labour is art, are going to be more interesting, exciting, and inspiring than the other pieces of art we could be seeing, such as those created by Pete Burns. If we have to make a choice, between one and the other, we will usually go for the highest score.

The piece you are about to see was published on a blog post through WordPress on the 2nd of October of 2016 (Pinheiro, 2016).

MATHEMATICAL FALLACIES V: APPLICATION OF INEQUALITIES

PROOF. Integrating by parts we have

$$f(x) - \frac{1}{b-a} \int_a^b f(t)\,dt = \frac{1}{b-a} \int_a^b p(x,t)f'(t)\,dt, \qquad (2.2)$$

for all $x \in [a, b]$, where

$$p(x,t) := \begin{cases} t-a, & \text{if } t \in [a, x], \\ t-b, & \text{if } t \in (x, b]. \end{cases}$$

(Dragomir, 1997)

In (Kouba, 2000), we read:

The formula for the method of integration by parts is given by

$$\int u \, dv = uv - \int v \, du \ .$$

It is a constant in Sever's writings, this thing of using the *same tricks*. Here, he also does the multiplication of dimensions/variables. Notice that we go from one to two, since we have **(x,t)**, not only **(x)**, in the newly created formula, which now brings **p(x,t)**.

Let's first multiply his original expression in this piece by **(b-a)**, shall we?

(b-a) f(x) – Int(a,b) f(t)dt – Int(a,b) p(x,t)f'(t)dt, where **p(x,t)** might be either **(t-a)** or **(t-b)**, and therefore where **p(x,t)** will have derivative **dt**.

Observing the formula we got from (Kouba, 2000), the one above, we would have to call **u** either **(b-a)** or **f(x)**, and **f(x)** will be either **u** or **v**, depending on what we decide to do with **u**.

If **u = (b-a)**, **du** should be **0**, since both **a** and **b** are constants. If **u = f(x)**, **du** is **f'(x)dx**.

If **du** is **0**, we get only the first two terms we see in the formula we got from (Kouba, 2000), and therefore only one integral. This is not what Sever got. In this case, we must have **u = f(x)** and **du = f'(x) dx**.

If **u** is **f(x)**, we need to find a multiplication in Sever's result in order to find out who **v** is, something without an integral. **f(x)** appears in isolation in his result, so that we must have **v = 1**. If **v** is **1**, **dv** is **0**, and we would have only one integral in the result, not two, so that applying the technique of integrating by parts and getting the result Sever claims to have gotten seems to be an impossible task.

If **dv** is **f'(t)dt**, and **u** is **p(x,t)**, this admitting that we can apply a one-dimensional theory to a two-dimensional expression, we would have to have, as a result, according to the formula for integration by parts, which is what Sever claims to be using, **p(x,t)f(t)- int(f(t)p'(x,t)d(x,t))**.

Now, the definition of the function **p(x,t)** is again *sui generis*. To get around the fact that he should have a two-dimensional function, Dragomir actually forces it now. We now have only one variable from those two. Oh, well, it is either **t-a** or **t-b** then. Once more, as said on the other blog post, we cannot really use **x** in the way he does.

Assuming, however, that we can, we now have **(t-a)f(t)- int(f(t)dt)** or **(t-b)f(t)- int(f(t)dt)**, assuming we can do this replacement.

Now let's go back to the formula for integration by parts, which we got from (Kouba, 2000), and try to find the missing term on it. If what we now have is **uv – int (vdu)**, **int (udv)** is **int ((t-a)f'(t) dt)** or **int ((t-b)f'(t) dt)**. That is perhaps, following his reasoning, the same as **int (p(x,t)f'(t)dt)** or **int (p(x,t) f'x(dx))**. In any hypothesis, it does not look possible to find anything compatible in his writing. Perhaps if **(b-a)f(x)** were the same as **(t-a)f(t)**, but **t** is not a constant according to his definition, **t** runs over part of the real interval, so that they could not possibly be the same. **(b-a)** would be the same as **(t-a)** only if **b** were the same as **t**, but **b** is a constant and **t** is a variable. Now, using his division into two intervals, **x** could be the same as **t** for that piece. The problem is that **t** is still a variable, not a constant, and one of its possible values is the own **x**.

The analogous procedure works for definite integration by parts, so

$$\int_a^b u \, d v = [u \, v]_a^b - \int_a^b v \, d u,$$

where $[f]_a^b = f(b) - f(a)$.

(Weisstein, 2016)

Even if we use the extract above, we cannot get the expected result because **[uv] from a to b** would be **uv(b) – uv(a)**. If we now have **(t-a)f(t)- int(f(t)dt)** or **(t-b)f(t)- int(f(t)dt)**, **uv(b)** could be the same as **(b-a)f(b)** and **uv(a)** could be the same as **(a-b)f(a)**, what then would give us **uv(b)-uv(a)=(b-a)(f(b)+f(a))**.

Well, with this, we have to invalidate whatever follows, and therefore the entire paper.

Now, because, just like with all Dragomir does, this paper leads to tons of others, all the others will also be invalidated.

One of those is (Masjed-Jamei, 2016).

References

02091977BX. (2014). Pete Burns You Spin Me Round. Retrieved 25 Dec 2016 from https://www.youtube.com/watch?v=dvqO126QJ3U

Queen Official. (2008). Queen – Bohemian Rhapsody (Official Video). Retrieved 25 Dec 2016 from https://www.youtube.com/watch?v=fJ9rUzIMcZQ&list=PLZYzh1QhBgMark6rrridAXQbozFrlxc12

Victoria University. (2016). Professor Silvestru Sever Dragomir. Retrieved 25 Dec 2016 from https://www.vu.edu.au/contact-us/silvestru-sever-dragomir

Dragomir, S. S., & Wang, S. (1997). An Inequality of Ostrowski-Griiss' Type and Its Applications to the Estimation of Error Bounds for Some Special Means and for Some Numerical Quadrature Rules. *Computers and Mathematical Applications*, *33*(11), 15–20.

Kouba, D. (2000). The Method of Integration by Parts. Retrieved October 5, 2016, fromhttps://www.math.ucdavis.edu/~kouba/CalcTwoDIRECTORY/intbypartsdirectory/IntByParts.html

Masjed-Jamei, M., Omey, E., & Dragomir, S. S. (2016). A Main Class of Integral Inequalities with Applications. *Mathematical Modelling and Analysis*, *21*(4), 569–584. Retrieved fromhttps://www.researchgate.net/publication/304717804_A_Main_Class_of_Integral_Inequalities_with_Applications?pli=1&loginT=mnWYF1-e4fd72x074TYEzf85mI4h-CjocFiXVWAqgKDb4VfP9RAbFw&uid=nxhS3fG8nldSXX9XFDkFMkdXT8GqA1rPVCBl&cp=re378_rpb_nreq_p2000&ch=reg

Weisstein, E. W. (2016). Integration by Parts. Retrieved 25 Dec 2016 from http://mathworld.wolfram.com/IntegrationbyParts.html

Sever had some funny assertions. Perhaps he had a lot of humour. I was asked to revise his book (he asked me), and I gladly accepted, since I was really good at it. The book was formed of papers of fellows as well as paper of his. I would have some serious observations sometimes, like things that should have been much better written in my opinion.

He would then turn to me and say that what mattered, in a mathematical proof, was repeating what was written in the theorem both in the beginning and in the end of it, for example.

The blog post you see next (Pinheiro, 2016) is more about the names of things. Mathematics has very specific names for its elements, and we don't usually like when people swap the name of things.

MATHEMATICAL FALLACIES IV: TRIANGLE INEQUALITY

Triangle Inequality

EXPLORE THIS TOPIC IN
The MathWorld Classroom

Let x and y be vectors. Then the triangle inequality is given by

$$||x| - |y|| \leq |x + y| \leq |x| + |y|. \tag{1}$$

Equivalently, for complex numbers z_1 and z_2:

$$||z_1| - |z_2|| \leq |z_1 + z_2| \leq |z_1| + |z_2|. \tag{2}$$

Geometrically, the right-hand part of the triangle inequality states that the sum of the lengths of any two sides of a triangle is greater than the length of the remaining side.

A generalization is

$$\left| \sum_{k=1}^{n} a_k \right| \leq \sum_{k=1}^{n} |a_k|. \tag{3}$$

(Weisstein, 2016)

Now, see how it looks like in the *modern application* of the concept:

The equality case holds in (1) if and only if there exists a constant $\lambda \in \mathbb{K}$ such that $x = \lambda y$.

In 1985 the author [5] (see also [24]) established the following refinement of (1):

$$\|x\| \, \|y\| \geq |\langle x, y \rangle - \langle x, e \rangle \langle e, y \rangle| + |\langle x, e \rangle \langle e, y \rangle| \geq |\langle x, y \rangle| \tag{2}$$

for any $x, y, e \in H$ with $\|e\| = 1$.

Using the triangle inequality for modulus we have

$$|\langle x, y \rangle - \langle x, e \rangle \langle e, y \rangle| \geq |\langle x, e \rangle \langle e, y \rangle| - |\langle x, y \rangle|$$

(Dragomir, 2016)

Any words about it?

Yes, I am also short of those.

Now, some people invented the term *reverse triangle inequality* for this situation, so that we could have said that we used the reverse triangle inequality instead. See:

A Proof of the Reverse Triangle Inequality

Let's suppose without loss of generality that $|x|$ is no smaller than $|y|$. (Otherwise we just interchange the roles of x and y.) Thus we have to show that

$$\|x\| - \|y\| \leq \|x - y\| \quad (*)$$

This follows directly from the triangle inequality itself if we write x as

$$x = x - y + y$$

and think of it as

$$x = (x-y) + y$$

Taking norms and applying the triangle inequality gives

$$\|x\| = \|x - y + y\| \leq \|x - y\| + \|y\|$$

which implies $(*)$

Fine print, your comments, more links, Peter Alfeld, PA1UM

[15-Mar-1998]

(Alfred, 1998)

References

Pinheiro, M. R. (2016). Mathematical Fallacies IV: Triangle Inequality. Retrieved 26th Dec 2016 from https://drmarciapinheiro.wordpress.com/2016/10/02/triangle-inequality/

Alfred, P. (1998). A Proof of the Reverse Triangle Inequality. Retrieved 26th December 2016 from https://www.math.utah.edu/~pa/math/equations/proof.html

Dragomir, S. S. (2016). Some Vector Inequalities for two Operators in Hilbert Spaces with Applications. Retrieved 26th December 2016 from https://www.degruyter.com/downloadpdf/j/ausm.2016.8.issue-1/ausm-2016-0005/ausm-2016-0005.xml?origin=publication_detail&ev=pub_int_prw_xdl&msrp=Lhdk96sd5TtGS5-db6RQuxDF7qCzFio1RYdYZ0b2w2aRDpjIziqpHOrjurkTcacsK7REXZznTbACvujCkEY3u3nWLR8VXfqdxf2bFc-0ZD8.co9ptng0uoqEGzMyBnpTzKeDIc6DtGkrnVrPuwpf3x3wD8QJMW4a7Q3voeHeaYUFCmOYXyNdnw6iMyWywIQXCQ.7zvTIg27vBQPny7TWOE0ugxRjYjxJBB4eQneFrhQiR6ke-RX_TJ5GPbARs3wSUyqTS8Cay6H74UHbtLHMT8aeA

Weisstein, E. W. (2016). Triangle Inequality. Retrieved 26 Dec 2016 from http://mathworld.wolfram.com/TriangleInequality.html

We had noticed other recurrent problems with Sever's writing whilst reading his work in 2001. Again, we did mention that to him in person, also argued, and again he had that perhaps cheeky way of dealing with it. One of them is seen below in a paper we recently published.

In (Pinheiro, 2008), we see:

From Dragomir and Fitzpatrick [1], we copy the method below to determine the left bound for the inequality:

— Consider the midpoint situation in the definition of convexity:

$$f(0.5(x + y)) \leq 0.5f(x) + 0.5f(y);$$

— (Problem 1) Now re-write x as $x := \lambda x + (1 - \lambda)y$ and y as $y := (1 - \lambda)x + \lambda y$;

— Integrate all over the variable λ, which ranges from 0 to 1 via definition, in the interval of definition;

— Then they would claim the result was achieved. We will follow their reasoning, but we are already making public we oppose to all of it. See:

$$f(0.5(x + y)) \leq 0.5f(\lambda x + (1 - \lambda)y) + 0.5f((1 - \lambda)x + \lambda y)$$

$$\Longleftrightarrow \int_0^1 f(0.5(x + y))d\lambda \leq 0.5 \int_0^1 f(\lambda x + (1 - \lambda)y)d\lambda$$

$$+ 0.5 \int_0^1 f((1 - \lambda)x + \lambda y)d\lambda$$

(Problem 2)

$$\Longleftrightarrow f(0.5(x + y)) \leq \frac{1}{y - x} \int_0^1 f(x)dx.$$

However, the problems with this proof are so extraordinary that it is unavoidable to think how it was thought to be plausible before. First of all, we had two free variables at the beginning, in the definition of convexity. Basically, they behave as belonging to different dimensions, that is, we hold a definition using the function twice, same function, what makes the inequality live at $\Re^2$, rather than $\Re$. The variables are made to live freely, therefore. It cannot be the case that, contrary to the intended definition, we simply make ties between them, so that there is exclusion of possibilities, without ever mentioning that or providing a way of reverting that process it is all absurd. In (Problem 1), that is the issue. Basically, there would be a trial of making a single dimensional variable become an object in a 3-dimensional space (variables are: x, λ, y, first instance). That is perfectly insane, if not accompanied of a very good explanation and a way to revert the process. Just to provide the reader with simplest reasoning of all, with that assignment, once λ is chosen, and x is chosen, we are left with a single choice for y. However, before that, y would be freely chosen even after the first two choices. It is missing at least a correspondence function which allows us to progress and also revert the process to the initial variable at the end the symbols used are applied in computation, *Maple*. However, even there, the process must be reversible.

In (Problem 2), the result is actually zero, for it is missing noticing it is $x - y$ in one of them, but it is $y - x$ in the other, and that obviously makes it all untrue, providing enough support to the thesis that the step noticed in (Problem 1) should never have been dared.

After so many years, all we have to add is that we should probably have written one dimensional variable instead of a single dimensional variable.

This multiplication of variables, just like the multiplication of bread by Jesus, is something frequent in Sever's writings.

The second reference that appears in this references list should be the [1] we referred to in the extract.

In another piece of the paper, we refer to the proof elaborated by Dragomir and McAndrew as excellent, and, because of that, we deduce another result, which we deem to be sound.

Unfortunately, later on, we would find the mistake also in the proof we deemed excellent in this piece.

We will first exhibit the extract where we call the proof excellent, and then the extract where we find a mistake in it, and therefore the extract where we nullify the result that followed, our own result.

In McAndrew and Dragomir [3], we find a better proof for the left bound for HH. There, it reads (call it PROOF Z):

Proof. Assume f is differentiable and convex in (a, b), then, for each $x, y \in (a, b)$, one has the inequality

$$f(x) - f(y) \geq (x - y)f'(y).$$

Using the property of modulus, we then get

$$f(x) - f(y) - (x - y)f'(y) = |f(x) - f(y) - (x - y)f'(y)|$$
$$\geq ||f(x) - f(y)| - |x - y||f'(y)||,$$

for each $x, y \in (a, b)$. Now choose $y = 0.5(a + b)$ and replace y on previous inequality to get

$$f(x) - f(0.5(a + b)) - (x - 0.5(a + b))f'(0.5(a + b))$$
$$\geq ||f(x) - f(0.5(a + b))| - |x - 0.5(a + b)||f'(0.5(a + b))||,$$

for any $x \in (a, b)$. Now, integrating that all in $[a, b]$:

$$\int_a^b [f(x) - f(0.5(a + b)) - (x - 0.5(a + b))f'(0.5(a + b))]dx$$

$$\geq \int_a^b ||f(x) - f(0.5(a + b))| - |x - 0.5(a + b)||f'(0.5(a + b))||dx,$$

$$\int_a^b f(x)dx - \int_a^b f(0.5(a+b))dx$$

$$- \int_a^b xf'(0.5(a+b))dx + 0.5(a+b)f'(0.5(a+b))(b-a)$$

$$\geq \int_a^b ||f(x) - f(0.5(a+b))| - |x - 0.5(a+b)||f'(0.5(a+b))||dx.$$

Dividing all by $(b-a)$, we then get:

$$\frac{1}{b-a}\int_a^b f(x)dx - f(0.5(a+b))$$

$$- \frac{1}{b-a}f'(0.5(a+b))\int_a^b xdx + 0.5(a+b)f'(0.5(a+b))$$

$$\geq \frac{1}{b-a}\int_a^b ||f(x) - f(0.5(a+b))| - |x - 0.5(a+b)||f'(0.5(a+b))||dx.$$

or

$$\frac{1}{b-a}\int_a^b f(x)dx - f(0.5(a+b))$$

$$\geq \frac{1}{b-a}\int_a^b ||f(x) - f(0.5(a+b))| - |x - 0.5(a+b)||f'(0.5(a+b))||dx \geq 0.$$

This closes our proof for good. $\square$

Remark 2. The above proof is excellent, but it makes us notice that any value, rather than $0.5(a+b)$, could have been used as y. What that gives us is that any value of the function may be used in the HH-inequality, without harm to result.

And now comes the part where we realized our own mistake when saying that the proof presented by McAndrew and Dragomir had no mistake, the proof we called PROOFZ.

(Pinheiro, 2011) brings a really sad abstract, where we say that the results will have to be nullified. That really hurts, and it must hurt anyone who does research: Nothing worse than having to nullify our own results. Yet, it is our duty, I would say ethical duty, to disclose the finding as soon as possible, since each, and every, result in Science will lead to tons of others: In the same way we followed Dragomir and McAndrew in that paper, tons of other people will follow us. Mistake in Science is something that probably grows in an exponential way.

The best way to avoid mistakes, I find, is realistically always writing our papers as if we are teaching others for the first time what the theory behind what we are doing is. The shorter the paper, the easier for people to check its results, and the more we explain about those, the more chances we have to reduce the amount of wrong results, and therefore waste in Science.

International Journal of Pure and Applied Mathematics

Volume 71   No. 1   2011, 31-39

ON S-CONVEXITY AND OUR PAPER WITH IJPAM

I.M.R. Pinheiro

P.O. Box 12396, A'Beckett st, Melbourne
Victoria, 8006, AUSTRALIA

Abstract: In this paper, we review the results of the paper H-H inequality for S-convex functions, published in the prestigious academic vehicle IJPAM. Substantial part of those results will have to be nullified. Most of the time, the mistakes have been inherited from other authors' work, so that we are also providing argumentation for the nullification of those authors' results in this paper in an indirect way.

See why it is a mistake and what is a mistake in PROOFZ:

Pointed problems: We have deduced this result because of the proof presented in the paper of McAndrew and Dragomir (see [4]), proof that appears by the page 567 of [3]. However, such a proof is wrong and contains basic mistakes, some of which have been pointed by us throughout time (see our remarks on other results attained by Dragomir et al.). Notice, for instance, that, at the beginning of the proof located at the page 567 of [3], Dragomir et al. make use of a severely well known consequence of the definition of derivative of a function. y, there, is a variable, and the theorem can only make sense if y is a variable. Unfortunately, by the seventh line of our reproduction of that proof, Dragomir et al. replace y with a constant value, $0.5(a + b)$, without presenting any sound mathematical justification for such a move, what is obviously unacceptable.

Also, one must notice that the origin of all is a line where who is greater than who matters, for if $\mathbf{x} < \mathbf{y}$ and $\mathbf{f(x)} < \mathbf{f(y)}$ you have one thing, but if things are with a $>$ instead, we have another.

If such a thing is relevant to the results we get, we could not have integral from $\mathbf{a}$ to $\mathbf{b}$: We would have to stop before the $\mathbf{y}$ we chose instead. See (Pinheiro, 2011):

writing about deriving the function generically first then about applying the derivative function to that point, we have to adapt all to the x in the formula as well, trivially, what has to mean that x does not belong to (a, b) anymore, but to $(a, 0.5(a + b))$ instead, at most, because there is an assumption that $x \neq y$ (otherwise the ratio would not be defined) and $y > x$ in the derivative formula. Notwithstanding, if we assume what we have just suggested, the resulting inequality is not what we would like it to be.

One more from the same source and about the same source:

not correct either. In visiting the third page of [6], we find a series of basic mistakes. From the first and second lines, already, departing from inequality (2.2): x and y are variables appearing together in a simple sum, which has only x and y as addends. With Dragomir et al.'s suggestion, we actually get $x = ta + (1 - t)b$ and $y = tb + (1 - t)a$, what leads to $x + y = a + b$, therefore to a single possible evaluation of the inequality (2.2), rather than a set of them (variables), being it all unacceptable mathematically from this point onwards. Notice that one new variable is inserted when Dragomir et al. supposedly simply 'apply' the definition of s_2-convexity to the now 'constant' argument. Such has been observed before by us in their work, and is fully unacceptable mathematically, leading to the nullification of our previously 'claimed-to-be' left bound for this inequality. Therefore, this result must be **entirely nullified** as well.

References

Pinheiro, M. R. (2008). H-H Inequality for S-convex Functions. International Journal of Pure and Applied Mathematics, 44(4), pp. 563-579

Dragomir, S. S., Fitzpatrick, S. (1999). The Hadamard Inequalities for S-convex Functions in the Second Sense. Demonstratio Mathematica, 27(4)

Pinheiro, M. R. (2011). On S-convexity and Our Paper with IJPAM. International Journal of Pure and Applied Mathematics, 71(1), pp. 31-39

Nonclassical Logicians proud themselves on the fact that what they write contravene some basic law of Classical Logic, so that it is as if they are proud of being revolutionary. The funny thinking they have is that they changed all but forgot to adapt the notion of Entailment to what they do. See the paper that follows.

Mathematics Letters
2016; 2(4): 28-31
http://www.sciencepublishinggroup.com/j/ml
doi: 10.11648/j.ml.20160204.11

An Issue with the Concept of Entailment

Marcia R. Pinheiro

Department of Mathematics and Philosophy, IICSE University, Wilmington, USA

Email address:
drmarciapinheiro@gmail.com

To cite this article:
Marcia R. Pinheiro. An Issue with the Concept of Entailment. *Mathematics Letters.* Vol. 2, No. 4, 2016, pp. 28-31.
doi: 10.11648/j.ml.20160204.11

Received: September 4, 2016; **Accepted:** October 12, 2016; **Published:** October 21, 2016

Abstract: Entailment is an interesting sigmatoid: It should mean one thing, but it means another, just for starters. When used in Mathematics, it is usually with the sense of saying that something is definitely true. That would be the use in Classical Logic then. When used in Logic, it became something else. Now it was about how the logical system, which can be any nonclassical one, could be making a proposition become true or false. The major issue we found in 2000, when learning from the own nonclassicists what they do, was that they talk about Nonclassical Logic, therefore a way of thinking that is not Cartesian, yet they stick to the notion of entailment we use in Mathematics, and therefore to the Classical Logic ways. We here discuss exactly this.

Keywords: Logical System, Logic, Nonclassical, Classical, Entailment, Implication

1. Introduction

From (Pinheiro, 2016), we read the following:

> 1.1.5 It is also standard to define two notions of validity. The first is *semantic*. A valid inference is one that *preserves truth*, in a certain sense. Specifically, every interpretation (that is, crudely, a way of assigning truth values) that makes all the premises true makes the conclusion true. We use the metalinguistic symbol '⊨' for this. What distinguishes different logics is the different notions of interpretation they employ.

Figure 1. Two notions of validity, Priest.

When we have nothing to the left side followed by the symbol ⊨, and then a proposition, what is usually meant is that we have a tautology to the right, that is, something that is true in all interpretations that are allowed in the system (Johnstone, 1987).

Some authors, such as Johnstone (1987), say that that is the same as having {} ⊨, and then the proposition, what then mean that we need no propositions to support a tautology.

An interpretation is a set of truth-values being assigned to the propositions under consideration. If we have the empty set to the left side, we obviously do not have to assign truth-values to any proposition, so that the antecedent of the entailment becomes automatically true (vacuum rule perhaps).

The implication is that if we have that all the interpretations that make what is to the left side, which we are calling antecedent (Introduction, 2004), true also make what is to the right side true, and that is what we are calling consequent (Introduction, 2004), then whatever is to left entails whatever is to the right.

If the consequent is always true for any interpretation we choose, what has to be true for tautologies, and there is no interpretation to be considered in terms of antecedent, we should probably have a conflict, but the decision is that in vacuum, OK. That probably comes from the fact that for us to falsify an implication in Classical Logic (CL), there is only one way: antecedent is true, consequent is false. If we guarantee that the consequent is true, then the implication will always be true, like regardless of the antecedent.

Nonclassical Logic has been created by the revolutionary: They wanted CL to be used ALSO in real life. In real life, people have doubts, so that something might be true, and false at the same time for at least some amount of time. True, and false were not enough in this case.

Nonclassical Logic came to pervert the rules from Classical Logic, therefore, to basically promote revolution in it: Why are you Cartesian in what regards gender? We have men, women, and middle-sex. When we want to assign a truth-value to a sentence of the type *I am a woman*, we will be in trouble if the person is from the middle-sex. Accept a maybe instead of a true or false, and it is all fixed.

Another scholar found a fourth way to go, and there was another system. Fuzzy Logic (Rouse, 2016) came up with infinitely many, and so on.

What we find most interesting however is that they want to use, according to Priest (2001), the same notion of entailment that we use in Mathematics, and therefore in CL. Do half the work, and say you did the same, basically.

We think we are sure they should at least have thought for longer about it, so perhaps they should have produced a paper on the topic, some book chapters, and whatever else to show that they have thought about it.

Intuitively, the concept of entailment would have to adapt to the logical system we have at hand, so that if we have a three-valued system, so say that we have the truth-values A, B, and C, and rules that perhaps agglutinate to the left side of the entailment, so say two values to the left give the same result in the end if the right side keeps its value, so say A, and B to one side, and A to the other would still give us OK or A, we would have to at least write that in the definition of entailment, it seems. In this case, instead of whatever interpretation makes the left side become true, we should have whatever interpretation makes the left side become true or middle-value, like to the least.

In this paper, we try to discuss this interesting concept of entailment, and propose that nonclassicists think more about the topic or expose what they have already discussed in a way for us to understand that they have already thought enough about it.

legal limitation. Since the logical system has rules, we can take those to be laws, and then say that if *A* entails *B*, then *A* provokes the appearance of a limitation for *B*.

If nothing appears to the left side, we understand there are no limitations, and that is when we say we have a tautology to the right side of the symbol ⊨, so that it is all making a lot of sense so far.

If this is the actual sense of *entails*, then the nonclassicists could be right in using it for their nonclassical systems without adapting or changing anything, but they then would probably have to stop saying that whatever makes the left side true would have to make the right side true.

The right way of putting it would probably be: If the left side does not bring any opposition to whatever is on the right side, then we have an entailment.

Now we can accept that any contradiction, and nothing would be the same, since it would not impose any limitation to the right side. Up to this date, however, we thought that there was a bit of confusion, since contradictions would not be true in CL.

If the left side said that *A* was zero, so say *A=0*, then the right side could not bring *A=2*, we assume.

There is a bit of confusion in the literature, however. Allan (2010) let us know that

Consider the following sentences:

(1a) No students laughed.
(1b) No students laughed loudly.

2. Development

So, if we had only one proposition, say *P: x belongs to the reals*, and we wanted to know if that entailed that *Q: x+2=5 => x=3*, we would have to play the following game: Assume that $v(P) = 1$. Now we know that *x* does belong to the reals. With this, we know that $v(Q) = 1$ as well. Therefore, $P \models Q$.

If instead we had *P: x belongs to the interval (7,10)*, we would know that *Q* would be true in the same way because both antecedent, and consequent of the implication would be false, what gives us a true implication, so that we would have $v(P)=1 => v(Q)=1$, and therefore $P \models Q$.

That is counter-intuitive in all. Perhaps we should revise the definition of entailment even in terms of Classical Logic.

Some people have discussed semantic intersection/connection in terms of entailment, and even implication (Mares, 1998) at waste, however, and nothing that could add seems to have been found.

It makes some sense to believe that if by making the left side true we get the right side true, then we have an entailment.

Entailment is defined in the following way (Harper, 2001):

mid-14c., "convert (an estate) into 'fee tail' (*feudum talliatum*)," from *en-* (1) "make" + *taile* "legal limitation," especially of inheritance, ruling who succeeds in ownership and preventing the property from being sold off, from Anglo-French *taile*, Old French *taille* "allot, cut to shape," from Late Latin *taliare* "to split" (see *tailor* (n.)). Sense of "have consequences" is 1829, via the notion of "inseparable connection." Related: Entailed; entailing; entailment.

Figure 2. Etymology of the word Entailment.

We then understand that it should mean the creation of a

(2a) Every philosopher smokes.
(2b) Every philosopher smokes heavily.

Whereas Example (1a) entails Example (1b), this is not the case with Examples (2a) and (2b). On the other hand, Example (2b) does entail Example (2a), whereas Example (1b) does not entail Example (1a). The important property to note here is that in the (b) sentences, the verb phrases pick out a smaller set of entities than the (a) sentences do (those who laugh loudly constitute a subset of those who laugh, and the set of heavy smokers is a subset of the set of smokers). So, the truth value of a sentence with *No students* in its subject position is unaffected if the denotation of its verb phrase (VP) is reduced. At the same time, the truth value of a sentence with *every philosopher* as subject remains unaffected whenever the denotations of its VP is extended.

Figure 3. Encyclopedia, Entails, definition.

In this case, the source claims that (1b) does not entail (1a). (1b) says that no students laughed loudly, and (1a) says that no students laughed. If no students laughed loudly, we are not saying that no student laughed, quite trivially, but (1a) would be a restriction on (1b), a limitation, like we would have reduced the group of students that laughed even further, so that, in our point of view, that would have been a limitation. (1b) does not limit (1a), it is the opposite, so that we here would think that all is agreeing with the just

proposed new definition for entailment, since (1a) does entail (1b) according to the source.

It also claims that (2a) does not entail (2b). (2b) says that every philosopher smokes heavily. (2a) says that every philosopher smokes. (2b) would clearly be a restriction on (2a), therefore a limitation, so that (2b) does entail (2a). On the other hand, (2a) would not entail (2b) if we consider our new definition, since (2a) is not limiting (2b), it is the opposite.

The source seems to have the same understanding we just acquired here: Limitations are what entails, a reduction in the domain provoked by the left side of the relationship we analyse.

From (Stanford, 2016), comes the following extract:

Logical Entailment

We say that a sentence φ *logically entails* a sentence ψ (written φ |= ψ) if and only if every truth assignment that satisfies φ also satisfies ψ. More generally, we say that a set of sentences Δ *logically entails* a sentence ψ (written Δ |= ψ) if and only if every truth assignment that satisfies all of the sentences in Δ also satisfies ψ.

For example, the sentence *p* logically entails the sentence (*p ∨ q*). Since a disjunction is true whenever one of its disjuncts is true, then (*p ∨ q*) must be true whenever *p* is true. On the other hand, the sentence *p* does *not* logically entail (*p ∧ q*). A conjunction is true if and only if *both* of its conjuncts are true, and *q* may be false. Of course, any

inseparable connection as an alternative, and he mentions that this sense appeared in the year of 1829.

If the sense is *to have consequences*, then the left side of the entailment would have as a consequence the right side, what is then compatible with the definition we find in Stanford (2016), and is also compatible with the definition we find in (Allan, 2010), considering the examples he there gives.

Now, this thing of being true to one side leading to all being true to the other or not imposing a situation in which the other side would not be true, is actually connected to the only way to falsify an implication in CL, which is antecedent true when consequent isn't.

We are only using two possible truth-values here, true, and false, and therefore we are obviously using CL.

Notwithstanding, Fuzzy Logic would have an infinity of possible truth-values, and we perhaps would need to consider those when talking about entailment inside of that system. That is actually the point of this paper.

As we go from CL to nonclassical systems, we should probably also find new definitions for entailment, and not only for validity.

Hajek (2002) says that

The standard set of truth degrees is the real interval [0, 1] with its natural ordering ≤ (1 standing for absolute truth, 0 for absolute falsity); but one can work with different domains, finite or infinite, linearly or partially ordered.

its conjuncts are true, and *q* may be false. Of course, any set of sentences containing both *p* and *q* does logically entail ($p \wedge q$).

Note that the relationship of logical entailment is a purely logical one. Even if the premises of a problem do not logically entail the conclusion, this does not mean that the conclusion is necessarily false, even if the premises are true. It just means that it is *possible* that the conclusion is false.

Once again, consider the case of ($p \wedge q$). Although *p* does not logically entail this sentence, it is *possible* that both *p* and *q* are true and, therefore, ($p \wedge q$) is true. However, the logical entailment does not hold because it is also possible that *q* is false and, therefore, ($p \wedge q$) is false.

Note also that logical entailment is not the same as logical equivalence. The sentence *p* logically entails ($p \vee q$), but ($p \vee q$) does not logically entail *p*. Logical entailment is not analogous to arithmetic equality; it is closer to arithmetic inequality.

This definition matches that of Dr. Priest (2001), and fellows, but is completely different from the definition we see in (Allan, 2010).

So, they say that *p* entails *p or q*. We may think that we can replace their definition of entailment with an implication: *p -> p or q*. Notwithstanding, seeing things from closer, we would get antecedent false validating the implication as well, but they reduce it all to antecedent true with consequent true, so that we only have one case of the three allowed cases in the implication when it comes to their entailment: It is something apart.

Harper (2001) gives us the sense *have consequences* or

domains, finite or infinite, linearly or partially ordered. Truth functions of connectives have to behave classically on the extremal values 0,1.

He is talking about Fuzzy Logic.

We read (Priest, 2001), and found only one definition of entailment, which is the one we present here. It is possible that that is wrong because if you are changing your truth values, and instead of two you now have even infinitely many, you would have to change the way you think of entailment for it all to make sense, like true, and false was for CL.

We could then have, in the case of Fuzzy Logic, that whatever makes the left side receive a truth-value between 0.5, and 1 makes the right side receive a truth-value between 0.5, and 1 instead.

We feel that what they shouldn't have changed they did change, which is the concept of Completeness (Pinheiro, 2016), and what they should have changed, they didn't change, which is the concept of entailment.

3. Conclusion

Nonclassicists would have to come up with articles to at least justify their choices in terms of the definition of entailment: If they change truth-values, that should provoke a change also in terms of the definition of entailment, since that definition comes attached to truth-values. The definition says that whatever makes the antecedent true would have to make the consequent true for us to have an entailment.

It is possible that they would have to present a new definition of entailment for each nonclassical system that be not bivalent.

Here we have the opposite to what we had when we talked about Completeness (Pinheiro, 2016): We should be changing everything that has been created for CL.

Not entering details on how Fuzzy Logic has been used or defined so far, just talking about truth-values, it could be that we would have to say that entailment in Fuzzy Logic is only justified if both antecedent, and consequent are marked with a value that is between *0.5*, and *1*.

It seems that the best translation for the symbol |= is *have consequences* or *inseparable connection*. Perhaps the best way to word it would be *has as a consequence*.

In this way, if $X \models Y$, *X* has, as a consequence, *Y*.

This is something different from the implication because, in terms of CL, for instance, if the antecedent of the implication is evaluated as false, and the consequent as true, the implication is true, but, with the entailment, we only accept true to both sides as a way to validate the relationship.

References

[1] Pinheiro, M. R. (2016). Completeness. *IOSR - Jornal of Mathematics*. *12*(5), 34–37. Retrieved from https://www.researchgate.net/publication/307122839_Completeness

[2] Johnstone, P. T. (1987). *Notes on Logic and Set Theory*. Cambridge University Press. Retrieved from

A&hl=en&sa=X&ved=0ahUKEwj1vsrXsfDOAhXG7hoKHSguAsw4KBDoAQgaMAA#v=onepage&q=logical%20entailment%20tautology&f=false

[3] Rouse, M. (2016). Fuzzy Logic. Retrieved September 27, 2016, from http://whatis.techtarget.com/definition/fuzzy-logic

[4] Priest, G. (2001). *An Introduction to Non-Classical Logic*. Cambridge University Press.

[5] Mares, E. (1998). Relevance Logic. Retrieved October 15, 2016, from http://plato.stanford.edu/entries/logic-relevance

[6] Harper, D. (2001). Entail. Retrieved September 3, 2016, from http://www.etymonline.com/index.php?search=Entailment

[7] Allan, K. (2010). *Concise Encyclopedia of Semantics*. Elsevier. Retrieved from https://books.google.com.au/books?id=3_1snsgmqU8C&pg=PA561&lpg=PA561&dq=does+not+entail+example+logic&source=bl&ots=jAX-weH0ml&sig=CPfESWbKBv7_sO8ABAz1H5xREDM&hl=en&sa=X&ved=0ahUKEwiX1LSOn_POAhXCE5QKHXf2Afo4ChDoAQgeMAE#v=onepage&q=does%20not%20entail%20example%20logic&f=false

[8] Stanford Logic Group. (2016). Logical Properties and Relationships. Retrieved September 3, 2016, from http://logic.stanford.edu/intrologic/notes/chapter_03.html

[9] Hajek, P. (2002). Fuzzy Logic. Retrieved September 4, 2016, from http://plato.stanford.edu/entries/logic-fuzzy

[10] Introduction to Logic. (2004). Retrieved September 4, 2016, from http://philosophy.lander.edu/logic/conditional.html

As another point, it seems that their notiion of Completeness is a little equivocated. See:

IOSR Journal of Mathematics (IOSR-JM)
e-ISSN: 2278-5728, p-ISSN: 2319-765X. Volume 12, Issue 5 Ver. I (Sep. - Oct.2016), PP 34-37
www.iosrjournals.org

Completeness

Dr.Pinheiro[1]

Abstract: *In this paper, we investigate the concept of completeness. We studied the concept whilst still attending college, but that wasfrom a mathematical perspective. In 2000, we got to have contact with The Logicians' understanding of the concept through the hands of one of the most important modern icons of Philosophy, Dr. Graham Priest. We recently mentioned his ways of applying the concept, and that was in our last paper with the APM journal. There seems to be a bit of discrepancy. Because of that, it is worth studying the subtleties involved. It seems that reserved words should not be recreated in meaning, so that if there is any chance The Logicians' completeness does not coincide with The Mathematicians' completeness, the sense that last appeared should be dropped in favour of coherence.*
Keywords: Logic, priest, completeness, mathematics, gödel

I.   Introduction

The completeness question for the first order predicate calculus was stated precisely and in print for the firsttime in 1928 by Hilbert and Ackermann in their text Grundzüge der theoretischenLogik [3], a text with whichGödel would have been quite familiar. [6]
The question Hilbert and Ackermann pose is whether a certain explicitly given axiom system for the first orderpredicate calculus "...is complete in the sense that from it all logical formulas that are correct for each domain of individuals can be derived..." ([4], p. 48).

(Pinheiro. 2016)

This is the mathematical definition we have learned at school: The axiomatic system is complete if we can derive all formulas that are correct for each allowed domain from it.

Basically, mathematicians would have a lot of difficulties because their minds would go: Give me a formula that is correct and I will then prove that it can be derived from our axiomatic system.

The person who listens to them would then say: Here is formula X.

They would then say: Here is the proof that it can be derived from our axiomatic system.

Now. how can we be sure that any other randomly chosen formula that is correct could be derived from the same system?

How can we know how many formulas are correct without writing them down?

If we consider a really small set of Classical Logic symbols. so say $=>$. Λ. and V. and a few letters. so say

A, B, and C, we already have. considering a special arrangement:

Choice of the first element: 3 (letters)
Choice of the second element: 3 (symbols)
Choice of the third element: 3 (letters)

3^3 or 27 possible formulas. We must take away from the counting the situation where we have the same letter before and after the symbol. so that we must take away three for each letter and each symbol, what gives us 9 situations to take away. We must also take away the situation involving .and. or .or. and permutation. so that we have combination of three by twos. 3. to take away here for each one of them. therefore 6 to take away. In total. we must take away 15 cases. what still gives us 12 possible cases.

As we add complexity to the logical formula. figures grow quite a lot: If we simply add an implication before the previously considered formula and a letter, with all we had before being put inside of brackets. we get three choices for every antecedent and therefore 3x12=36 possible cases.

The mathematician would then go crazy: How can I possibly check on all correct formulas?

Priest and his fellows seem to have come up with a solution to all the mathematicians' nightmares in what regards this topic: You simply depart from a generic couple of letters organized in a way to have a symbol in the middle and you then assume that the formula is correct. After that, you see if it is possible going back to the

---

[1]IICSE University. drmarciapinheiro@gmail.com

DOI: 10.9790/5728-1205013437   www.iosrjournals.org   34 | Page

value of the letter via allowed inferential rules (allowed in that particular logical system we study). You would then have to just rewrite things in the proper order to show that it is possible to derive that particular formula from the axiomatic system.

In the example we have just provided, our efforts would resume to checking three formulas $(A \Rightarrow B, A \wedge B,$ and $A \vee B)$.

Wow! That is miraculous, right?

We went from 12 to 3 in a single life: Oh, gosh!

So, is this the miracle of all times or there is something wrong going on there?

That was the question in my mind in that 2000 as I watched Priest skilfully running his fingers through the board and doing his weaving: He would go upwards from the lowest levels of the thing he drew, which was like a tree. Trees I was used to: That came from Systems Analysis.

The resulting unit, the formula, let's say, was at the bottom.

The main question that emerges in our mind is: Can we really cover all formulas like that, like all formulas that could be correct and made out of the allowed symbols for this system?

In this paper, we shall endeavour to try to answer such a question.

II. Development

We observe that in (Pinheiro, 2016) we proved that there is no counter-example to the claim *Arithmetic is complete*. We actually only have circumstantial evidence on the contrary.

To prove that a system is incomplete, it suffices that we find a counter-example to the claim that it is complete, so that it suffices that we find one formula that cannot be derived from the system we have at hand, basically.

To prove that a system is complete, however, we would have to prove that ALL allowed formulas that are true can be derived from the system we have.

Whilst logicians like Priest are using the sigmatoids (Pinheiro, 2015)*completeness* and *soundness* at waste and creating numerous logical systems from day to night, mathematicians would have the rights to feel at least suspicious: Is that something that we are allowed to do?

The book Priest showed us in that 2000 (Priest, 2001) brought the following:

1.11.5 COMPLETENESS LEMMA: Let b be an open complete branch of a tableau. Let v be the interpretation induced by b. Then:

if A is on b, $v(A) = 1$

if $\neg A$ is on b, $v(A) = 0$

1.11.6 COMPLETENESS THEOREM: For finite Σ, if $\Sigma \models A$ then $\Sigma \vdash A$.

1.3.3 Let Σ be any set of formulas (the premises); then A (the conclusion) is a *semantic consequence* of Σ ($\Sigma \models A$) iff there is no interpretation that makes all the members of Σ true and A false, that is, every interpretation that makes all the members of Σ true makes A true. '$\Sigma \not\models A$' means that it is not the case that $\Sigma \models A$.

1.3.4 A is a *logical truth* (*tautology*) ($\models A$) iff it is a semantic consequence of the empty set of premises ($\phi \models A$), that is, every interpretation makes A true.

The detail that most matters here is that completeness, for Dr. Priest, means that whatever makes the set of premises to the left side true also makes the set of premises to the right true, but, in Mathematics, in what comes to completeness, we would need to guarantee is the opposite direction: For every premise that is true to the right, it is possible to find a set of premises that are true from which we can derive them to the left. The concepts are therefore incompatible. The fact that the set of facts X makes whatever is to the left and to the right true does not mean that it is not possible to find another true assertion to the right that does not find derivation on whatever is

DOI: 10.9790/5728-1205013437   www.iosrjournals.org   35 | Page available to the left. Therefore they have not solved Gödel's main problem. This is at most a proof of coherence of some sort and, more than likely, could not be called completeness, since that is a reserved name, basically.

1.1.5 It is also standard to define two notions of validity. The first is *semantic*. A valid inference is one that *preserves truth*, in a certain sense. Specifically, every interpretation (that is, crudely, a way of assigning truth values) that makes all the premises true makes the conclusion true. We use the metalinguistic symbol '⊨' for this. What distinguishes different logics is the different notions of interpretation they employ.

(Priest, 2001)

Semantic validity appeared because people like Dr. Corcoran (Corcoran, 1999) were looking for some relationship that matters between antecedent and consequent in Classical Logic: They wanted to have some connection in terms of meaning between one thing and another. The name *Semantic Validity* is ALSO equivocated therefore. There was an expectation and a search that mattered to those people: They were looking for an almost spiritual connection between left and right. If the definition of Semantic Validity is the one we just presented here, and this is the definition Dr. Priest uses, we cannot guarantee any semantic attachment because we can, for instance, have a tautology to the right side and therefore any interpretation that makes the left side true will make the right side also true. That does not imply any semantic connection whatsoever.

1.1.6 The second notion of validity is *proof-theoretic*. Validity is defined in terms of some purely formal procedure (that is, one that makes reference only to the symbols of the inference). We use the metalinguistic symbol '⊢' for this notion of validity. In our case, this procedure will (mainly) be one employing tableaux. What distinguish different logics here are the different tableau procedures employed.

(Priest, 2001)

Here we are just talking about the rules of the system and mathematicians would therefore not have any problems with this.

The Completeness Theorem here exhibited states that, for a finite set of premises, *whenever those and the right side are true* (observe that they use *if..then*, not the CL symbol)*there is a proof of whatever appears to the right side that can be made out of these premises*.

In this case, say we have A -> B, C to the left side and C to the right.

Assuming that A->B is true in CL is saying that it is never the case that v(A)=1 and v(B)=0. Assuming that C is true in CL is saying that v(C)=1. With this, we obviously get the right side to have value 1 or to be true, what would then tell us that there is a deduction originating in the left side that leads to the right side.

On the other hand, assume that D is also true and we have no other premises to count on to the left side, so say D is a tautology, like regardless of the values we use from this system, D is always true. That will obviously mean that D is true for every interpretation of sigma. Assume that A, B, C, and D have nothing in common. Now, we do not have completeness in the mathematical sense in this system, which involves only A->B, and C, but we do have completeness in the logical sense according to the definition we hold.

What we are actually checking here is if whenever the truth of a set of premises does not lead to the falsity of another we are sure to have a proof of the latter inside of that set of premises, all being just a matter of organizing things properly. We are then only worried about the case in which we have 1 -> 1, what is odd. In this case, we will guarantee that there is a proof of whatever comes to the right. 0 -> 1 would also be OK, but we are not guaranteeing this case.

Ex-Falso is excluded because if we have that both A->B and ~(A->B) are true, that is not acceptable in CL. They cannot both be true at the same time. Notwithstanding, if we had that to the left side and assumed it was all true we would have explosion and therefore ALSO whatever is in the right side. The truth of whatever is to the left is still leading to the truth of whatever is to the right. Yet, we should not have a proof of whatever is to the right in this way because, first of all, we cannot have both being true at the same time. Otherwise anything and

DOI: 10.9790/5728-1205013437

everything we say would be provable in the system: It suffices getting the thing we want to prove and its negation.

As another point, Priest talks about many-valued logics, what would mean that it would be possible to have both cases validated with no conflict if the logical system chosen were Special K, let's say: v(A->B)=1 and v(~(A->B))=1. The truth of those could therefore imply the truth of the right side, so say it is A->B. Now, how do we actually prove the right side?

A proof could imply the application of more than one rule, not just one, as a start of conversation. Is that a proof of A->B or just a fact?

Would we have to consider their definition of completeness inside of each logical system, according to their rules? We get the impression that they use CL to interpret their axiom in what regards completeness instead.

In the same sense, if all is implied, we don't seem to have a special proof of whatever is to the right side. It is only the application of a rule.

Any contradiction would be a proof in this case. We can create never ending ones by simply swapping one element.

This Completeness Theorem is either incomplete or equivocated.

What Priest does with his hands on the board, however, could be the actual thing: He comes from the basic formula, the most basic units of all, and he goes back to the root of those formulas, to the most elementary parts, so say A->B. We assume this premise is true, and therefore that v(A->B)=1 if we are in CL.

For that to be true, either B is true or A is false. If our universe is CL, provided we can evaluate all As and Bs as true or false, we are OK.

How do we evaluate a contradiction if we find it in place of A, however? Is that true or false? We would have to say it is impossible if we talk about CL, and therefore we are stuck with being unable to provide a truth-value to it.

When he moves his hands therefore, in the weaving thing, he is not proving completeness either, very unfortunately.

Oh, well, in this case, the most he could be doing is checking the syntax involved for coherence, we assume.

Proving completeness, already said Gödel somehow, is something really difficult to do, so that logicians should refrain from using this mathematical sigmatoid from now onwards, we reckon.

III.   Conclusion

The inadequate use of the mathematical sigmatoid completeness led us to get very confused about the teachings of Priest in 2000. The reason is quite obvious: Logicians ARE NOT proving completeness.

In that being a drama for mathematicians for almost one century (Pinheiro, 2016), it would be unlikely that the logicians had solved all in so simple moves, but, even so, given the amount of work they have produced, those from nonclassical logic, we must think of the possibility and therefore take it all very seriously.

Logic, as much as Mathematics, should never be a joke.

The sigmatoid should definitely be withdrawn from the logical jargon until more adequate theorems and lemmas are found, in case that ever happens.

What they meant, more than likely, was syntactic coherence at most, we think.

Truth be said, if they had really found ways to prove CL completeness, Gödel's incompleteness theorems would not be a problem anymore, so that we had plenty of material evidence to count on when we decided to challenge their statements.

When we go from words to symbols, things become quite dangerous, as said by us many times. The opposite move is much more guaranteed.

References

[1].   Pinheiro, M. R. (2015). Words for Science. *Indian Journal of Applied Research*, 5(5), 19–22. Retrieved from https://www.worldwidejournals.com/ijar/articles.php?val=NjQ0MQ==&b1=853&k=214

[2].   Pinheiro, M. R. (2016). Gödel and the Incompleteness of Arithmetic. *Advances in Pure Mathematics*, 6(8), 9. Retrieved from http://file.scirp.org/Html/4-5301134_68410.htm

[3].   Priest, G. (2001). *An Introduction to Non-Classical Logic*. Cambridge University Press.

[4].   Corcoran, J. (1999). Information-Theoretic Logic and Transformation-Theoretic Logic. *Fragments of Science*, 25–36. Retrieved from http://philpapers.org/archive/CORILA.pdf